Materials in Our World

FOSS
FULL OPTION SCIENCE SYSTEM

3rd Edition

Developed at the Lawrence Hall of Science, University of California, Berkeley
Published and Distributed by Delta Education

FOSS Lawrence Hall of Science Team
Larry Malone and Linda De Lucchi, FOSS Project Codirectors and Lead Developers
Kathy Long, FOSS Assessment Director; David Lippman, Project Specialist and Editor; Carol Sevilla, Publications Design
Coordinator; Rose Craig, Illustrator; Susan Stanley, Graphic Production; John Quick, Photographer; Trudihope Schlomowitz,
Barbara Clinton, Editors; DeSondra Ward, Office Manager
FOSS Curriculum Developers: Brian Campbell, Teri Dannenberg, Alan Gould, Susan Kaschner Jagoda, Ann Moriarty,
Jessica Penchos, Kimi Hosoume, Virginia Reid, Terry Shaw, Joanna Snyder, Erica Beck Spencer, Diana Velez, Natalie Yakushiji
FOSS Technology Developers: Susan Ketchner, Habiba Noor, Arzu Orgad
FOSS Multimedia Team: Kate Jordan, Multimedia Director; Nicole Medina, Senior Multimedia Producer;
Matthew Jacoby, Lead Programmer; Geoffrey Thomas, Multimedia Programmer and Designer; Chris Linden, Designer;
Chris Hamamoto, Designer; Dan Bluestein, Programmer; Roger Vang, Programmer; Christopher Cianciarulo, Programmer

Delta Education Team
Bonnie A. Piotrowski, FOSS Editorial Director
Project Team: Mathew Bacon, Jennifer Apt, Sandra Burke, Diane Gonciarz, Tom Guetling, Joann Hoy, Jacquelyn Lachance,
Lisa Lachance, John Prescott

Thank you to all FOSS Grades K-6 Trial Teachers
Heather Ballard, Wilson Elementary, Coppell, TX; Terra L. Barton, Harry McKillop Elementary, Melissa, TX; Rhonda Bernard,
Frances E. Norton Elementary, Allen, TX; Heather Callaghan, Olive Chapel Elementary, Apex, NC; Katie Cannon, Las Colinas
Elementary, Irving, TX; Kristy Cash, Wilson Elementary, Coppell, TX; Monica Coles, Swift Creek Elementary School, Raleigh,
NC; Melissa Cook-Airhart, Harry McKillop Elementary, Melissa, TX; Hillary P. Croissant, Harry McKillop Elementary, Melissa,
TX; Nancy Deveneau, Wilson Elementary, Coppell, TX; Marlana Dumas, Las Colinas Elementary, Irving, TX; Mary Evans, R.E.
Good Elementary School, Carrollton, TX; Jacquelyn Farley, Moss Haven Elementary, Dallas, TX; Corinna Ferrier, Oak Forest
Elementary, Humble, TX; Allison Fike, Wilson Elementary, Coppell, TX; Colleen Garvey, Farmington Woods Elementary, Cary,
NC; Erin Gibson, Las Colinas Elementary, Irving, TX; Dollie Green, Melissa Ridge Intermediate School, Melissa, TX; Amanda
Hill, Las Colinas Elementary, Irving, TX; Cindy Holder, Oak Forest Elementary, Humble, TX; Carol Kellum, Wallace Elementary,
Dallas, TX; Brittani Kern, Fox Road Elementary, Raleigh, NC; Jodi Lay, Lufkin Road Middle School, Apex, NC; Ana Martinez, RISD
Academy, Dallas, TX; Shaheen Mavani, Las Colinas Elementary, Irving, TX; Mary Linley McClendon, Math Science Technology
Magnet School, Richardson, TX; Adam McKay, Davis Drive Elementary, Cary, NC; Anne Mechler, J. Erik Jonsson Community
School, Dallas, TX; Shirley Diann Miller, The Rice School, Houston, TX; Anne Miller, J. Erik Jonsson Community School, Dallas, TX;
Keri Minier, Las Colinas Elementary, Irving, TX; Stephanie Renee Nance, T.H. Rogers Elementary, Houston, TX; Elizabeth Noble,
Las Colinas Elementary, Irving, TX; Sarah Peden, Aversboro Elementary School, Garner, NC; Carrie Prince, School at St. George
Place, Houston, TX; Marlaina Pritchard, Melissa Ridge Intermediate School, Melissa, TX; Alice Pujol, J. Erik Jonsson Community
School, Dallas, TX; Paul Rendon, Bentley Elementary, Oakland, CA; Janette Ridley, W.H. Wilson Elementary School, Coppell,
TX; Kristina (Crickett) Roberts, W.H. Wilson Elementary School, Coppell, TX; Heather Rogers, Wendell Creative Arts & Science
Magnet Elementary School, Wendell, NC; Megan Runion, Olive Chapel Elementary, Apex, NC; Christy Scheef, J. Erik Jonsson
Community School, Dallas, TX; Samrawit Shawl, T.H. Rogers School, Houston, TX; Ashley Stephenson, J. Erik Jonsson Community
School, Dallas, TX; Jolanta Stern, Browning Elementary School, Houston, TX; Gale Stimson, Bentley Elementary, Oakland, CA;
Cathryn Sutton, Wilson Elementary, Coppell, TX; Robin Taylor, Arapaho Classical Magnet, Richardson, TX; Michael C. Thomas,
Forest Lane Academy, Dallas, TX; Jomarga Thompkins, Lockhart Elementary, Houston, TX; Mary Timar, Madera Elementary,
Lake Forest, CA; Helena Tongkeamha, White Rock Elementary, Dallas, TX; Linda Trampe, J. Erik Jonsson Community School,
Dallas, TX; Charity VanHorn, Fred A. Olds Elementary, Raleigh, NC; Kathleen VanKeuren, Lufkin Road Middle School, Apex,
NC; Mary Margaret Waters, Frances E. Norton Elementary, Allen, TX; Lisa Webb, Madisonville Intermediate, Madisonville, TX;
Nancy White, Canyon Creek Elementary, Austin, TX; Barbara Yurick, Oak Forest Elementary, Humble, TX; Linda Zittel, Mira Vista
Elementary, Richmond, CA

Photo Credits: © Greg Nicholas/iStockphoto (cover); © John Quick; © Laurie Meyer; © Erica Beck Spencer

Published and Distributed by Delta Education, a member of the School Specialty Family
The FOSS program was developed in part with the support of the National Science Foundation grant nos. MDR-8751727 and
MDR-9150097. However, any opinions, findings, conclusions, statements, and recommendations expressed herein are those of
the authors and do not necessarily reflect the views of NSF. FOSSmap was developed in collaboration between the BEAR Center
at UC Berkeley and FOSS at the Lawrence Hall of Science. Score analysis is done through the BEAR Center Scoring Engine.

Materials in Our World — Teacher Toolkit, 1325299
Investigations Guide, 1391916
978-1-60902-667-7
Printing 3 – 7/2014
Webcrafters, Madison, WI

Materials in Our World

TABLE OF CONTENTS

Overview

Contents

INTRODUCTION

The **Materials in Our World Module** provides experiences that heighten primary students' awareness, curiosity, and understanding of the physical world and provides opportunities for young students to engage in scientific and engineering practices. Students observe and compare the properties of a variety of kinds of wood, paper, fabric, and earth materials. They discover what happens when they subject the materials to a number of tests and interactions. In this module, students will

- Observe and compare physical properties of different kinds of wood samples, using the senses.

- Observe and compare properties and structures of different kinds of paper and fabric.

- Observe how wood, paper, and fabric interact with water.

- Explore the technology of making wood products.

- Observe and describe how and where fabrics are used.

- Observe, describe, and mix earth materials with water to observe properties.

- Communicate observations made about different kinds of materials, orally and through drawings.

- Use knowledge of the properties of materials to create useful and/or aesthetic objects.

	Module Summary	Focus Questions
Inv. 1: Getting to Know Wood	Students work with five different wood samples to observe their properties. They begin with free exploration, go on a hunt for matching samples, drop water on the samples, and float them in basins. They test the wood to find out how many paper clips it takes to sink it, then organize their results by making a concrete graph.	Where does wood come from? What is made of wood? What happens when wood gets wet? How can you sink wood? How many paper clips does it take to sink wood?
Inv. 2: Changing Wood	Students use sandpaper to change the shape of wood. They compare sawdust and shavings and how they interact with water. They simulate the manufacture of two kinds of wood they observed in Investigation 1, particleboard and plywood.	How can you change the shape of wood? How are sawdust and shavings the same? How are sawdust and shavings different? How is particleboard made? How is plywood made?
Inv. 3: Getting to Know Paper	Students observe and compare the properties of ten kinds of paper and go on a hunt for matching samples. They compare how well the papers fold and which has the best surface for writing. They test papers for absorption, then soak the samples overnight. Students learn how to recycle paper by making new paper from old and crafting papier-mâché bowls.	What is made of paper? What makes paper good for writing? What makes paper easy to fold? What happens when water gets on paper? How can new paper be made from old paper? How can paper be made strong to form a bowl?
Inv. 4: Getting to Know Fabric	Students observe and compare the properties of ten kinds of fabric and search for different ways fabrics are used. They take apart fabrics to learn how they are woven from threads. Students investigate how fabrics interact with water. They soil, dye, wash, and dry fabrics to learn about temporary and permanent changes.	How are fabrics different? What is made of fabric? How is fabric made? What happens when water gets on fabric? What are some things that stain fabric? How are different kinds of fabric used?
Inv. 5: Earth Materials	Students observe rocks, sort them by properties, and describe changes when they put rocks in water. Students collect and describe soil samples and make mud paint to observe their colors. They freeze and melt water to observe phase change, and discuss water, rock, and other natural resources that can be conserved by reuse and recycling.	How are rocks different? What happens when soil gets wet? How can we change water? How can we conserve natural resources? What can we make from different materials?

Content	Reading	Assessment
• Wood has many observable properties. • Different kinds of wood come from different kinds of trees. Some kinds of wood are processed and made by people. • Wood floats in water but can be made to sink. • Wood absorbs water. • Trees are natural resources.	**Science Resources Book** "The Story of a Chair"	**Embedded Assessment** Teacher observation
• Wood can be changed by sanding and mixing with water. • Sawdust is tiny wood pieces that can be recycled. • Basic materials can be transformed into new materials (particleboard and plywood). • Mixtures form when materials are put together. • Liquid water in an open container evaporates.	**Science Resources Book** "Are You a Scientist?"	**Embedded Assessment** Teacher observation
• Paper has many observable properties. • People make paper from wood. • The properties of different papers determine their uses. • Some paper absorbs water; others do not. • Some paper changes when soaked in water. Some paper breaks down into small fibers.	**Science Resources Book** "The Story of a Box"	**Embedded Assessment** Teacher observation
• Fabric is a flexible material with many properties. • The properties of fabrics determine their uses. • Fabric can be made of woven threads. • Fabrics can absorb, transmit, or repel water. • Wet fabric dries when water evaporates, leaving the fabric unchanged. • Materials that interact with fabric may wash away or produce stains.	**Science Resources Book** "What Is Fabric Made From?" "How Are Fabrics Used?"	**Embedded Assessment** Teacher observation
• Rocks, soil, and water are earth materials. • Rocks can be compared, sorted, and described. • Soil can be described by its properties. • Water can change from solid to liquid (melt) with heat and from liquid to solid (freeze) with cold. • Land, air, water, and trees are natural resources. • People reuse and recycle to conserve resources.	**Science Resources Book** "How Are Rocks Different?" "How Are Rocks, Soil, and Water Used?" "Land, Air, and Water" "I Am Wood"	**Embedded Assessment** Teacher observation

FOSS CONCEPTUAL FRAMEWORK

In the last half decade, teaching and learning research has focused on learning progressions. The idea behind a learning progression is that **core ideas** in science are complex and wide-reaching, requiring years to develop fully—ideas such as the structure of matter or the relationship between the structure and function of organisms. From the age of awareness throughout life, matter and organisms are important to us. There are things we can and should understand about them in our primary school years, and progressively more complex and sophisticated things we should know about them as we gain experience and develop our cognitive abilities. When we can determine those logical progressions, we can develop meaningful and effective curriculum.

FOSS has elaborated learning progressions for core ideas in science for kindergarten through grade 8. Developing a learning progression involves identifying successively more sophisticated ways of thinking about a core idea over multiple years. "If mastery of a core idea in a science discipline is the ultimate educational destination, then well-designed learning progressions provide a map of the routes that can be taken to reach that destination" (National Research Council, *A Framework for K–12 Science Education*, 2012).

The FOSS modules are organized into three domains: physical science, earth science, and life science. Each domain is divided into two strands, as shown in the table below for the FOSS Elementary Program. Each strand represents a core idea in science and has a conceptual framework.

- matter; energy and change
- dynamic atmosphere; rocks and landforms
- structure and function; complex systems.

The sequence in each strand relates to the core ideas described in the national framework. Modules at the bottom of the table form the foundation in the primary grades. The core ideas develop in complexity as you proceed up the columns.

The FOSS learning progression information is displayed in several places: (1) the **module conceptual framework** (see page 7) represents the structure of scientific knowledge taught and assessed in a single module, and (2) the **content sequence** (pages 10–11) is a graphic and narrative description placing the single module into a K–8 strand content or learning progression.

TEACHING NOTE

FOSS has conceptual structure at the module and strand levels. The concepts are carefully selected and organized in a sequence that makes sense to students when presented as intended.

FOSS Elementary Module Sequences

	PHYSICAL SCIENCE		EARTH SCIENCE		LIFE SCIENCE	
	MATTER	**ENERGY AND CHANGE**	**DYNAMIC ATMOSPHERE**	**ROCKS AND LANDFORMS**	**STRUCTURE/ FUNCTION**	**COMPLEX SYSTEMS**
6	Mixtures and Solutions	Motion, Force, and Models	Weather on Earth	Sun, Moon, and Planets	Living Systems	
	Measuring Matter	Energy and Electromagnetism	Water	Soils, Rocks, and Landforms	Structures of Life	Environments
	Solids and Liquids	Balance and Motion	Air and Weather	Pebbles, Sand, and Silt	Plants and Animals	Insects and Plants
K	Materials in Our World		Trees and Weather		Animals Two by Two	

In addition to the science content development, every module provides opportunities for students to engage in and understand the importance of scientific practices, and many modules explore issues related to engineering practices and the use of natural resources.

Asking questions and defining problems

- Ask questions about objects, organisms, systems, and events in the natural and human-made world (science).

- Ask questions to define and clarify a problem, determine criteria for solutions, and identify constraints (engineering).

Planning and carrying out investigations

- Plan and conduct investigations in the laboratory and in the field to gather appropriate data (describe procedures, determine observations to record, decide which variables to control) or to gather data essential for specifying and testing engineering designs.

Analyzing and interpreting data

- Use a range of media (numbers, words, tables, graphs, images, diagrams, equations) to represent and organize observations (data) in order to identify significant features and patterns.

Developing and using models

- Use models to help develop explanations, make predictions, and analyze existing systems, and recognize strengths and limitations of proposed solutions to problems.

Using mathematics and computational thinking

- Use mathematics and computation to represent physical variables and their relationships and to draw conclusions.

Constructing explanations and designing solutions

- Construct logical explanations of phenomena, or propose solutions that incorporate current understanding or a model that represents it and is consistent with available evidence.

Engaging in argument from evidence

- Defend explanations, develop evidence based on data, examine one's own understanding in light of the evidence offered by others, and challenge peers while searching for explanations.

Obtaining, evaluating, and communicating information

- Communicate ideas and the results of inquiry—orally and in writing—with tables, diagrams, graphs, and equations —in collaboration with peers.

TEACHING NOTE

Crosscutting concepts, as identified by the national framework, bridge disciplinary core ideas and provide an organizational framework for connecting knowledge from different disciplines into a coherent and scientifically based view of the world. The **Materials in Our World Module** *addresses these crosscutting concepts: patterns; cause and effect; structure and function; and stability and change.*

BACKGROUND FOR THE CONCEPTUAL FRAMEWORK
in Materials in Our World

For thousands of years, humans have used natural fibers from both plant and animal sources to produce useful materials. The short, tough fibers of cellulose found in the stems and branches of plants are useful for making familiar materials that are found all over the world. Cellulose fibers can be teased apart by brute force and made into a pulp. When the pulp is spread out in a uniform layer, pressed flat, and dried, the resulting material is paper.

The first paper mill in the United States was built just over 300 years ago. It used discarded cotton and linen rags exclusively to manufacture paper. In the 1860s, when the demand for paper outstripped the supply, new techniques were developed to manufacture paper from wood. At the turn of the century, 60 percent of the paper made in America was derived from wood, with the rest coming from rags and reclaimed paper, and by 1930 nearly all paper came from virgin fiber sources. Since that time, people have been making advances in using paper fiber a second time, but there is still lots of room for improvement.

Long fibers of hair (wool from sheep, angora goats, alpacas, and the like) or single fibers from plants (cotton seed pods, hemp stems, and flax stems), twisted together, make useful, strong, manageable materials called thread and yarn. Threads and yarns can be manufactured into some of the most varied and important materials on Earth. Known variously as fabric, cloth, material, and textile, these flexible, durable, attractive, and shapeable materials provide protection, warmth, identity, shelter, tools, and beauty.

Rocks have also been important to humans in many ways for thousands of years. Rocks provided early people with shelter, weapons, and means for creating sparks to start fires for cooking and warmth. The ancient Egyptians built their pyramids out of rock, and the Greeks used stones called calculi for adding and subtracting. Hopi women used flat stones of different roughness to grind corn. Children during America's colonial days wrote on flat pieces of rock called slate. Astronauts collected rocks during their visits to the Moon. They brought these rocks back to scientists who hoped to unlock the secrets of the solar system by studying them.

Wood

The wide distribution of wood and its combination of properties make it an especially useful and versatile material. Because wood is most often obtained from trees, and trees come in thousands of different varieties around the world, there are thousands of different kinds of wood. Each kind of wood has observable properties such as color, grain pattern, and density (does it float or sink in water?). Some wood absorbs water, and other types of wood repel water. The properties of wood make it a useful natural resource for many purposes.

Wood production usually begins with the felling of a living tree rather than a dead one. Living trees are more abundant, the wood is in better condition, and they are easier to handle and less likely to split or break during harvest and transportation. Logs are hurried to the lumber mill and kept wet to prevent cracking and splitting until they are sawed.

Some kinds of wood are not from a single tree but are a product of modern technology—manufactured wood. Particleboard and plywood are two examples of manufactured wood. Particleboard was invented to take advantage of abundant waste wood fiber. Over the years, the process of making this material has been refined to produce a wood used in cabinetry and furniture as well as home construction. Plywood is made from thin sheets of wood glued together, with alternate layers of grain running in perpendicular directions. This construction produces a tremendously strong, large surface of wood, much more suitable for walls and floors than is conventional lumber.

CONCEPTUAL FRAMEWORK
Physical Science, Matter: Materials in Our World

Matter Has Structure

Concept A   Matter exists in three states (solid, liquid, and gas), which have observable properties.

Concept B   Matter has physical properties that can be observed and quantified. (Materials absorb water or repel water; materials float in water or sink in water.)

Matter Interacts

Concept B   Change of temperature can produce changes in physical state (water freezes, ice melts, water dries up).

Concept C   During physical interactions, substances form mixtures in which the interacting substances retain their original properties.

Paper

Paper is most often associated with some form of communication—paper as a medium for recording words or images: newspapers, books, letters, advertising flyers, drawings, photographs, and so forth. Paper was developed originally as a lightweight, reasonably durable, flat surface for painting and writing. Even the most cursory glance around today's environment will affirm that the original purpose has been fulfilled at a level certainly undreamed of by the pioneer papermakers.

The craft of papermaking is about 2000 years old. Paper is an example of a nonnative, or manufactured, material. People used native materials, such as stone, metal, wood, leather, and bone, long before they used manufactured materials, because manufactured materials require an accumulation of knowledge and the application of technology to produce them. The technology for paper production has advanced over the centuries. Today's paper mills make a huge variety of papers, ranging from delicate facial tissue and tracing paper to strong, durable kraft paper for shopping bags to heavy cardboard and tar paper for industrial use.

Paper falls into the category of renewable resources: the raw material to make paper is constantly being made in nature. If you need a new sheet of paper, you cut down a tree, reduce it to chips, smash and churn the bits into pulp, and make it into paper. If you need a new tree to make the next sheet of paper, you plant it, and nature in time delivers the tree. The false notion of a limitless supply of trees gave rise to the behavior of using paper once and tossing it out.

The forest resource is finite. The typical American uses 260 kilograms (kg) of paper each year, and to supply the Sunday paper to an eager public requires a virtual forest of trees—a whopping 500,000 trees each week. The next generations will have to be better educated than their parents, and their children better educated yet, about the value of fibers and the most effective ways to reuse them. Paper will be categorized with several other materials, as not only renewable but also reusable.

This module might give you and students a little bit of X-ray vision when you go out and look at trees You may be able to look through the bark to see the lumber, the sawdust, the bits of plywood, and the variety of papers that are the promise of a tree. On the other hand, when you look at a piece of wood or a sheet of paper, you will be able to see the legacy of the living tree.

Fabric

Fabric is also a manufactured material. The weaving of fabric is an ancient enterprise, dating back perhaps 5000 years. Natural fibers—probably hair from sheep and goats—were twisted into yarns and woven into a durable cloth, using a frame called a loom. The basic process of weaving threads together in two directions has not changed conceptually; it has just become more efficient and refined. A length of modern fabric might be 1 meter (m) or so wide and 100 m long. When it is wrapped around a piece of cardboard, ready for shipping to the fabric vendor, it is called a bolt of fabric. The long piece of fabric has threads that run the entire length of the bolt, 100 m or longer. These threads are the warp of the fabric.

The threads that run across the fabric are shorter, and they are woven over, under, over, under the threads of the warp. These are the woof of the fabric. It is extremely tedious to actually sinuate the woof over and under each thread of the warp, and long ago looms were invented that streamlined the process. If a weaver uses colored threads in the warp, the fabric produced can have interesting colors woven into it. If the weaver varies the number and arrangements of threads, intricate patterns can be woven into the fabric.

When only a few fabrics were available, each had a fairly well-defined set of applications. Wool made tough, weather-resistant, warm cloth, and linen made fine, lightweight, attractive frocks and trim items. Cotton made utilitarian items, and silk was used for regal finery. With the invention of synthetic fibers, the constellation of fabrics has expanded. Textile production is one of the most important industries worldwide and a wonderful subject for scientific investigation for primary students.

Rocks and Soil

Observing rocks and beginning to sort them into groups are the initial steps primary students take in their role as geologists studying earth materials. Students use these observations to make comparisons and to sort rocks into groups with similar properties. Students can put the rocks in water to enhance the color and pattern.

To an engineer, soil is any ground that can be dug up by earth-moving equipment and requires no blasting. To a primary student, soil is dirt, but they also come to know it as a mixture of different-size earth materials, such as gravel, sand, and silt. Soil also contains water and organic material called humus. Adding water to soil changes its properties temporarily and allows students to study the texture and colors of soil.

Matter Content Sequence

This table shows the five FOSS modules and courses that address the matter-content sequence for grades K–8. Running through the sequence are the two progressions—matter has structure, and matter interacts. The supporting elements in each module (somewhat abbreviated) are listed. The elements for the **Materials in Our World Module** are expanded to show how they fit into the sequence.

Module or course	MATTER	
	Matter has structure	**Matter interacts**
Chemical Interactions	• Matter is made of atoms. • Substances are defined by chemical formulas. • Elements are defined by unique atoms. • The properties of matter are determined by the kinds and behaviors of its atoms. • Atomic theory explains the conservation of matter.	• During chemical reactions, particles in reactants rearrange to form new products. • Energy transfer to/from the particles in a substance can result in phase change. • During dissolving, one substance is reduced to particles (solute), that are distributed uniformly throughout the particles of the other substance (solvent).
Mixtures and Solutions	• Solid matter can break into pieces too small to see. • Mass is conserved (not created or lost) during changes. • Properties can be used to identify substances (e.g., solubility). • Relative density can be used to seriate solutions of different concentrations.	• A mixture is two or more intermingled substances. • Dissolving occurs when one substance disappears in a second substance. • A chemical reaction occurs when substances mix, and new products result. • Melting is an interaction between one substance and heat.
Measuring Matter	• Properties of matter (solid, liquid, gas) can be described using measurement (length, mass, volume, temperature). • Measurement can be used to confirm that the whole is equal to its parts.	• Different substances change state (e.g., melt or freeze) at different temperatures. • Mass is conserved when objects or materials are mixed.
Solids and Liquids	• Common matter is solid, liquid, and gas. • Solid matter has definite shape. • Liquid matter has definite volume. • Gas matter has neither definite shape nor volume and expands to fill containers. • Intrinsic properties of matter can be used to organize objects (e.g., color, shape).	• Solids interact with water in various ways: float, sink, dissolve, swell, change. • Liquids interact with water in various ways: layer, mix, change color. • Substances change state (e.g., melt or freeze) when heated or cooled.
Materials in Our World		

Conceptual Framework

The **Materials in Our World Module** aligns with the NRC *Framework*. The module addresses these K–2 grade band endpoints described for core ideas from the national framework for physical sciences and for engineering, technology, and the application of science.

Physical Sciences
Core idea PS1: Matter and Its Interactions—How can one explain the structure, properties, and interactions of matter?

- *PS1.A: How do particles combine to form the variety of matter one observes?* [Different kinds of matter exist (e.g., wood, metal, water), and many of them can be either solid or liquid, depending on temperature. Matter can be described and classified by its observable properties, by its uses, and by whether it occurs naturally or is manufactured. Different properties are suited to different purposes. A great variety of objects can be built up from a small set of pieces. Objects or samples of a substance can be weighed, and their size can be described and measured. (Boundary: volume is introduced only for liquid measure.)]

- *PS1.B: How do substances combine or change (react) to make new substances? How does one characterize and explain these reactions and make predictions about them?* [Heating or cooling a substance may cause changes that can be observed. Sometimes these changes are reversible (e.g., melting and freezing) and sometimes they are not (e.g., baking a cake, burning fuel).

Matter has structure	Matter interacts
• Wood, paper, fabric, soil, and rock are examples of solid materials. • Solid objects are made of solid materials. • Solid objects have properties that can be described and compared. • The whole (object) can be broken down into smaller pieces.	• Wood, paper, and fabric can be changed by sanding, coloring, tearing, etc. • Common materials can be changed into new materials (papermaking, weaving, etc.). • Water can change to ice in a freezer, and ice can change to water in a room. • Water and soil mix to form mud.

Materials in Our World

> **TEACHING NOTE**
>
> *A Framework for K–12 Science Education* has two core ideas in engineering, technology, and applications of science.
>
> ETS1: Engineering Design
>
> ETS2: Links among Engineering, Technology, Science, and Society
>
> Core idea ETS1: How do engineers solve problems?
>
> ETS1.C: How can the various proposed design solutions be compared and improved? [Because there is always more than one possible solution to a problem, it is useful to compare designs, test them, and discuss their strengths and weaknesses.]
>
> Core idea ETS2: How are engineering, technology, science, and society interconnected?
>
> ETS2.B: How do science, engineering, and the technologies that result from them affect the ways in which people live? How do they affect the natural world? [People depend on various technologies in their lives. Every human-made product is designed by applying some knowledge of the natural world and is built by using materials derived from the natural world.]

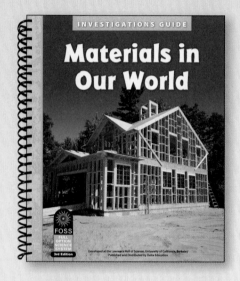

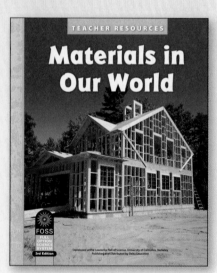

FOSS COMPONENTS

Teacher Toolkit

The *Teacher Toolkit* is the most important part of the FOSS Program. It is here that all the wisdom and experience contributed by hundreds of educators has been assembled. Everything we know about the content of the module, how to teach the subject, and the resources that will assist the effort are presented here. Each toolkit has three parts.

Investigations Guide. This spiral-bound document contains these chapters.

- Overview
- Materials
- Investigations (five in this module)
- Assessment

Teacher Resources. This collection of resources contains these chapters.

- FOSS Introduction
- Science Notebooks in Grades K–2
- Science-Centered Language Development
- Taking FOSS Outdoors
- FOSSweb and Technology
- Science Notebook Masters (for grades 1–6)
- Teacher Masters
- Assessment Masters

The chapters in *Teacher Resources* and the Spanish duplication masters can also be found on FOSSweb (www.FOSSweb.com).

FOSS Science Resources. This is a copy of the student book of readings that are integrated into the instruction.

Equipment Kit

The FOSS Program provides the materials needed for the investigations, including metric measuring tools, in sturdy, front-opening drawer-and-sleeve cabinets. Inside, you will find high-quality materials packaged for a class of 32 students. Consumable materials are supplied for two uses before you need to resupply. You will be asked to supply small quantities of common classroom items.

FOSS Science Resources Books

FOSS Science Resources: Materials in Our World is a book of original readings developed to accompany this module. The readings are referred to as articles in the *Investigations Guide*. Students read the articles in the book as they progress through the module. The articles cover a specific concept usually after that concept has been introduced in an active investigation.

The articles in *Science Resources* and the discussion questions provided in the *Investigations Guide* help students make connections to the science concepts introduced and explored during the active investigations. Concept development is most effective when students experience organisms, objects, and phenomena firsthand before engaging the concepts in text. The text and illustrations help make connections between what students experience concretely and the ideas that explain their observations.

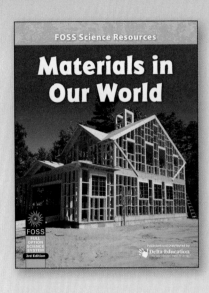

▶ **NOTE**
FOSS Science Resources: Materials in Our World is also provided as a big book in the kit.

▶ **NOTE**
To access all the teacher resources and to set up customized pages for using FOSS, log in to FOSSweb through an educator account.

FOSSweb and Technology

The FOSS website opens new horizons for educators, students, and families, in the classroom or at home. Each module has an interactive site where students and families can find instructional activities, interactive simulations and virtual investigations, and other resources. FOSSweb provides resources for materials management, general teaching tools for FOSS, purchasing links, contact information for the FOSS Project, and technical support. You do not need an account to view this general FOSS Program information. In addition to the general information, FOSSweb provides digital access to PDF versions of the *Teacher Resources* component of the *Teacher Toolkit* and digital-only resources that supplement the print and kit materials.

Additional resources are available to support FOSS teachers. With an educator account, you can customize your homepage, set up easy access to the digital components of the modules you teach, and create class pages for your students with access to tutorials and online assessments.

Ongoing Professional Development

The Lawrence Hall of Science and Delta Education are committed to supporting science educators with unrivaled teacher support, high-quality implementation, and continuous staff-development opportunities and resources. FOSS has a strong network of consultants who have rich and experienced backgrounds in diverse educational settings using FOSS. Find out about professional-development opportunities on FOSSweb.

FOSS INSTRUCTIONAL DESIGN

Each FOSS investigation follows a similar design to provide multiple exposures to science concepts. The design includes these pedagogies.

- Active investigation, including outdoor experiences
- Recording in science notebooks to answer the focus question
- Reading in *FOSS Science Resources*
- Assessment to monitor progress and motivate student reflection on learning

In practice, these components are seamlessly integrated into a continuum designed to maximize every student's opportunity to learn. An instructional sequence may move from one pedagogy to another and back again to ensure adequate coverage of a concept.

FOSS Investigation Organization

Modules are subdivided into **investigations** (five in this module). Investigations are further subdivided into 3–5 **parts**. Each part of each investigation is driven by a **focus question**. The focus question, usually presented as the part begins, signals the challenge to be met, mystery to be solved, or principle to be uncovered. The focus question guides students' actions and thinking and makes the learning goal of each part explicit for teachers. Each part concludes with students recording an answer to the focus question in their notebooks.

Investigation-specific **scientific background** information for the teacher is presented in each investigation chapter. The content discussion is divided into sections, each of which relates directly to one of the focus questions. This section ends with information about teaching and learning and a conceptual-flow diagram for the content.

The **Getting Ready** and **Guiding the Investigation** sections have several features that are flagged or presented in the sidebars. These include several icons to remind you when a particular pedagogical method is suggested, as well as concise bits of information in several categories.

Teaching notes appear in blue boxes in the sidebars. These notes comprise a second voice in the curriculum—an educative element. The first (traditional) voice is the message you deliver to students. It supports your work teaching students at all levels, from management to inquiry. The second educative voice, shared as a teaching note, is designed to help you understand the science content and pedagogical rationale at work behind the instructional scene.

FOCUS QUESTION

What is made of wood?

TEACHING NOTE

This focus question can be answered with a simple yes or no, but the question has power when students support their answers with evidence. Their answers should take the form "Yes, because _____."

The **safety** icon alerts you to a potential safety issue. It could relate to the use of a chemical substance, such as salt, requiring safety goggles, or the possibility of an allergic reaction when students use latex, legumes, or wheat.

The small-group **discussion** icon asks you to pause while students discuss data or construct explanations in their groups.

The **new-word** icon alerts you to a new vocabulary word or phrase that should be introduced thoughtfully. The new vocabulary should also be entered onto the word wall (or pocket chart). A complete list of the scientific vocabulary used in each investigation appears in the sidebar on the last page of the Background for the Teacher section.

The **vocabulary** icon indicates where students should review recently introduced vocabulary, often just before they will be answering the focus question or preparing for benchmark assessment.

The **recording** icon points out where students should make a science-notebook entry. Students record on prepared notebook sheets or, increasingly, on pages in their science notebooks.

The **reading** icon signals when the class should read a specific article in the *FOSS Science Resources* book, preferably during a reading period.

The **assessment** icon appears when there is an opportunity to assess student progress, by using embedded. Some of the embedded-assessment methods for grades K–2 include observation of students engaged in scientific practices, and review of a notebook entry (drawing or text).

The **outdoor** icon signals when to move the science learning experience into the schoolyard. It also helps you plan for selecting and preparing an outdoor site for a student activity.

The **engineering** icon indicates opportunities for addressing engineering practices—applying and using scientific knowledge. These opportunities include developing a solution to a problem, constructing and evaluating models, and using systems thinking.

The **EL Note** in the sidebar provides a specific strategy to assist English learners in developing science concepts. A discussion of strategies is in the Science-Centered Language Development chapter.

To help with pacing, you will see icons for **breakpoints**. Some breakpoints are essential, and others are optional.

EL NOTE

See the Science-Centered Language Development chapter for notebook-sharing strategies.

POSSIBLE BREAKPOINT

Active Investigation

Active investigation is a master pedagogy. Embedded within active learning are a number of pedagogical elements and practices that keep active investigation vigorous and productive. The enterprise of active investigation includes

- context: questioning and planning;
- activity: doing and observing;
- data management: recording/organizing/processing;
- analysis: discussing and writing explanations.

Context: questioning and planning. Active investigation requires focus. The context of an inquiry can be established with a focus question or challenge from you or, in some cases, from students. (What happens when wood gets wet?) At other times, students are asked to plan a method for investigation. This might start with a teacher demonstration or presentation. Then you challenge students to plan an investigation, such as to find out how many paper clips it takes to sink wood. In either case, the field available for thought and interaction is limited. This clarification of context and purpose results in a more productive investigation.

Activity: doing and observing. In the practice of science, scientists put things together and take things apart, observe systems and interactions, and conduct experiments. This is the core of science—active, firsthand experience with objects, organisms, materials, and systems in the natural world. In FOSS, students engage in the same processes. Students often conduct investigations in collaborative groups of four, with each student taking a role to contribute to the effort.

The active investigations in FOSS are cohesive, and build on each other and the readings to lead students to a comprehensive understanding of concepts. Through the investigations, students gather meaningful data.

Data management: recording/organizing/processing. Data accrue from observation, both direct (through the senses) and indirect (mediated by instrumentation). Data are the raw material from which scientific knowledge and meaning are synthesized. During and after work with materials, students record data in their science notebooks. Data recording is the first of several kinds of student writing.

Students then organize data so that they will be easier to think about. Tables allow efficient comparison. Organizing data in a sequence (time) or series (size) can reveal patterns. Students process some data into graphs, providing visual display of numerical data. They also organize data and process them in the science notebook.

Analysis: discussing and writing explanations. The most important part of an active investigation is extracting its meaning. This constructive process involves logic, discourse, and existing knowledge. Students share their explanations for phenomena, using evidence generated during the investigation to support their ideas. Students conclude the active investigation by writing a summary in their science notebooks of their learning as well as questions raised during the activity.

Science Notebooks

Research and best practice have led FOSS to place more emphasis on the student science notebook. Keeping a notebook helps students organize their observations and data, process their data, and maintain a record of their learning for future reference. The process of writing about their science experiences and communicating their thinking is a powerful learning device for students. The science-notebook entries stand as credible and useful expressions of learning. The artifacts in the notebooks form one of the core elements of the assessment system.

Reading in *FOSS Science Resources*

The *FOSS Science Resources* books emphasize expository articles and biographical sketches. FOSS suggests that the reading be completed during language-arts time. When language-arts skills and methods are embedded in content material that relates to the authentic experience students have had during the FOSS active learning sessions, students are interested, and they get more meaning from the text material.

Assessing Progress for Kindergarten

Assessment and teaching must be woven together to provide the greatest benefit to both the student and the teacher. Assessing young students is a process of planning what to assess, observing, questioning, and recording information about student learning for future reference. Observing students as they engage in the activity and as they share their notebook writing and drawings reveals their thinking and problem-solving abilities. Questioning probes for understanding. Both observing and questioning will give you information about what individual students can and can't do, and what they know or don't know. This information allows you to plan your instruction thoughtfully. For example, if you find students need more experience comparing the properties of materials, you can provide more time at a center that focuses on comparing similar materials or select extension activities that will continue to develop the ability to identify similarities and differences.

Use the techniques that work for you and your students and that fit with the overall kindergarten curriculum goals. The most detailed and reliable picture of students' growth emerges from information gathered using a variety of assessment strategies.

FOSS embedded assessments for kindergarten allow you and your students to monitor learning on a daily basis as you progress through the **Materials in Our World Module**. You will find suggestions for what to assess in the Getting Ready section of each part of each investigation. For example, here is the Getting Ready step for Part 1 of the first investigation.

13. **Plan assessment for Part 1**

There are six objectives that can be assessed at any time during any part of this investigation.

What to Look For

- *Students ask questions.*

- *Students plan and conduct simple investigations.*

- *Students use senses to observe materials and objects.*

- *Students record and organize observations.*

- *Students communicate observations orally and in their notebooks with words and drawings.*

- *Students incorporate new vocabulary.*

Here are specific content objectives to observe in this part.

- *Wood has observable properties and can be described by those properties.*

- *Wood is a resource that comes from trees.*

Here are the specific scientific practices to observe in this part.

- *Students compare properties of wood (organize observations).*

- *Students sort wood by properties (organize observations).*

Taking FOSS Outdoors

FOSS throws open the classroom door and proclaims the entire school campus to be the science classroom. The true value of science knowledge is its usefulness in the real world and not just in the classroom. Taking regular excursions into the immediate outdoor environment has many benefits. First of all, it provides opportunities for students to apply things they learned in the classroom to novel situations. When students are able to transfer knowledge of scientific principles to natural systems, they experience a sense of accomplishment.

In addition to transfer and application, students can learn things outdoors that they are not able to learn indoors. The most important object of inquiry outdoors is the outdoors itself. To today's youth, the outdoors is something to pass through as quickly as possible to get to the next human-managed place. For many, engagement with the outdoors and natural systems must be intentional, at least at first. With repeated visits to familiar outdoor learning environments, students may first develop comfort in the outdoors, and then a desire to embrace and understand natural systems.

The last part of most investigations is an outdoor experience. Venturing out will require courage the first time or two you mount an outdoor expedition. It will confuse students as they struggle to find the right behavior that is a compromise between classroom rigor and diligence and the freedom of recreation. With persistence, you will reap rewards. You will be pleased to see students' comportment develop into proper field-study habits, and you might be amazed by the transformation of students with behavior issues in the classroom who become your insightful observers and leaders in the schoolyard environment.

Teaching outdoors is the same as teaching indoors—except for the space. You need to manage the same four core elements of teaching: time, space, materials, and students. Because of the different space, new management procedures are required. Students can get farther away. Materials have to be transported. The space has to be defined and honored. Time has to be budgeted for getting to, moving around in, and returning from the outdoor study site. All these and more issues and solutions are discussed in the Taking FOSS Outdoors chapter in *Teacher Resources*.

FOSS is very enthusiastic about this dimension of the program and looks forward to hearing about your experience using the schoolyard as a logical extension of your classroom.

Science-Centered Language Development

The FOSS active investigations, science notebooks, *FOSS Science Resources* articles, and formative assessments provide rich contexts in which students develop and exercise thinking and communication. These elements are essential for effective instruction in both science and language arts—students experience the natural world in real and authentic ways and use language to inquire, process information, and communicate their thinking about scientific phenomena. FOSS refers to this development of language process and skills within the context of science as science-centered language development.

In the Science-Centered Language Development chapter in *Teacher Resources*, we explore the intersection of science and language and the implications for effective science teaching and language development. We identify best practices in language-arts instruction that support science learning and examine how learning science content and engaging in scientific practices support language development.

Language plays two crucial roles in science learning: (1) it facilitates the communication of conceptual and procedural knowledge, questions, and propositions, and (2) it mediates thinking—a process necessary for understanding. For students, language development is intimately involved in their learning about the natural world. Science provides a real and engaging context for developing literacy, and language-arts skills and strategies support conceptual development and scientific practices. For example, the skills and strategies used for enhancing reading comprehension, writing expository text, and exercising oral discourse are applied when students are recording their observations, making sense of science content, and communicating their ideas. Students' use of language improves when they discuss (speak and listen, as in the Wrap-Up/Warm-Up activities), write, and read about the concepts explored in each investigation.

There are many ways to integrate language into science investigations. The most effective integration depends on the type of investigation, the experience of students, the language skills and needs of students, and the language objectives that you deem important at the time. The Science-Centered Language Development chapter is a library of resources and strategies for you to use. The chapter describes how literacy strategies are integrated purposefully into the FOSS investigations, gives suggestions for additional literacy strategies that both enhance students' learning in science and develop or exercise English-language literacy skills, and develops science vocabulary with scaffolding strategies for supporting all learners. The last section covers language-development strategies specifically for English learners.

TEACHING NOTE

Embedded even deeper in the FOSS pedagogical practice is a bolder philosophical stance. Because language arts commands the greatest amount of the instructional day's time, FOSS has devoted a lot of creative energy defining and exploring the relationship between science learning and the development of language-arts skills. FOSS elucidates its position in the Science Centered-Language Development chapter.

FOSSWEB AND TECHNOLOGY

FOSS is committed to providing a rich, accessible technology experience for all FOSS users. FOSSweb is the Internet access to FOSS digital resources. It provides enrichment for students and support for teachers, administrators, and families who are actively involved in implementing and enjoying FOSS materials. Here are brief descriptions of selected resources to help you get started with FOSS technology.

Technology to Engage Students at School and at Home

Multimedia activities. The multimedia simulations and activities were designed to support students' learning. They include virtual investigations and student tutorials that you can use to support students who have difficulties with the materials or who have been absent.

FOSS Science Resources. The student reading book is available as an audio book on FOSSweb, accessible at school or at home. In addition, as premium content, *FOSS Science Resources* is available as an eBook. The eBook supports a range of font sizes and can be projected for guided reading with the whole class as needed.

Home/school connection. Each module includes a letter to families, providing an overview of the goals and objectives of the module. Most investigations have a home/school activity that provides science experiences to connect the classroom experiences with students' lives outside of school. These connections are available in print in the *Teacher Resources* binder and on FOSSweb.

Student media library. A variety of media enhance students' learning. Formats include photos, videos, an audio version of each student book, and frequently asked science questions. These resources are also available to students when they log in with a student account.

Recommended books and websites. FOSS has reviewed print books and digital resources that are appropriate for students and prepared a list of these media resources.

Class pages. Teachers with a FOSSweb account can easily set up class pages with notes and assignments for each class. Students and families can access this class information online.

> ▶ **NOTE**
> The FOSS digital resources are available online at FOSSweb. You can always access the most up-to-date technology information, including help and troubleshooting, on FOSSweb. See the FOSSweb and Technology chapter for a complete list of these resources.

Technology to Support Teachers

Teacher-preparation video. The video presents information to help you prepare for a module, including detailed investigation information, equipment setup and use, safety, and what students do and learn through each part of the investigation.

Science-notebook masters and teacher masters. All notebook masters (grades 1–6) and teacher masters used in the modules are available digitally on FOSSweb for downloading and for projection during class. These sheets are available in English and Spanish.

Focus questions. The focus questions for each investigation are formatted for classroom projection and for printing onto labels that students can glue into their science notebooks.

Equipment photo cards. The cards provide labeled photos of equipment supplied in each FOSS kit.

Materials Safety Data Sheets (MSDS). These sheets have information from materials manufacturers on handling and disposal of materials.

Teacher Resources chapters. FOSSweb provides PDF files of all chapters from the *Teacher Resources* binder.

- FOSS Introduction
- Science Notebooks
- Science-Centered Language Development
- Taking FOSS Outdoors
- FOSSweb and Technology

Streaming video. Some video clips are part of the instruction in the investigation, and others extend concepts presented in a module.

Resources by investigation. This digital listing provides online links to notebook sheets, assessment and teacher masters, and multimedia for each investigation of a module, for projection in the classroom.

Interactive-whiteboard resources. You can use these slide shows and other resources with an interactive whiteboard.

Investigations eGuide. The eGuide is the complete FOSS *Investigations Guide* component of the *Teacher Toolkit* in an electronic web-based format, allowing access from any Internet-enabled computer.

▶ **NOTE**
The Spanish masters are available only on FOSSweb.

UNIVERSAL DESIGN FOR LEARNING

The roots of FOSS extend back to the mid–1970s and the Science Activities for the Visually Impaired and Science Enrichment for Learners with Physical Handicaps projects (SAVI/SELPH). As those special-education science programs expanded into fully integrated settings in the 1980s, hands-on science proved to be a powerful medium for bringing all students together. The subject matter is universally interesting, and the joy and satisfaction of discovery are shared by everyone. Active science by itself provides part of the solution to full inclusion.

Many years later, FOSS began a collaboration with educators and researchers at the Center for Applied Special Technology (CAST), where principles of Universal Design for Learning (UDL) had been developed and applied. FOSS continues to learn from our colleagues about ways to use new media and technologies to improve instruction. Here are the UDL principles.

Principle 1. Provide multiple means of representation. Give learners various ways to acquire information and knowledge.

Principle 2. Provide multiple means of action and expression. Offer students alternatives for demonstrating what they know.

Principle 3. Provide multiple means of engagement. Help learners get interested, be challenged, and stay motivated.

The FOSS Program has been designed to maximize the science-learning opportunities for students with special needs and students from culturally and linguistically diverse origins. FOSS is rooted in a 30-year tradition of multisensory science education and informed by recent research on UDL. Strategies found effective with students with special needs and students who are learning English are incorporated into the materials and procedures used with all students.

English Learners

The FOSS multisensory program provides a rich laboratory for language development for English learners. The program uses a variety of techniques to make science concepts clear and concrete, including modeling, visuals, and active investigations in small groups at centers. Key vocabulary is usually developed within an activity context with frequent opportunities for interaction and discussion between teacher and student and among students. This provides practice and application of the new vocabulary. Instruction is guided and scaffolded through carefully designed lesson plans, and students are supported throughout. The learning is active and engaging for all students, including English learners.

Science vocabulary is introduced in authentic contexts while students engage in active learning. Strategies for helping all primary students read, write, speak, and listen are described in the Science-Centered Language Development chapter. There is a section on science-vocabulary development with scaffolding strategies for supporting English learners. These strategies are essential for English learners, and they are good teaching strategies for all learners.

Differentiated Instruction

FOSS instruction allows students to express their understanding through a variety of modalities. Each student has multiple opportunities to demonstrate his or her strengths and needs. The challenge is then to provide appropriate follow-up experiences for each student. For some students, appropriate experience might mean more time with the active investigations. For other students, it might mean more experience building explanations of the science concepts orally or in writing or drawing. For some students, it might mean making vocabulary more explicit through new concrete experiences or through reading to students. For some students, it may be scaffolding their thinking through graphic organizers. For other students, it might be designing individual projects or small-group investigations. For some students, it might be more opportunities for experiencing science outside the classroom in more natural, outdoor environments.

There are several possible strategies for providing differentiated instruction. The FOSS Program provides tools and strategies so that you know what students are thinking throughout the module. Based on that knowledge, read through the extension activities for experiences that might be appropriate for students who need additional practice with the basic concepts as well as those ready for more advanced projects. Interdisciplinary extensions are listed at the end of each investigation. Use these ideas to meet the individual needs and interests of your students.

ORGANIZING THE CLASSROOM

Students in primary grades are usually most comfortable working as individuals with materials. The abilities to share, take turns, and learn by contributing to a group goal are developing but are not reliable as learning strategies all the time. Because of this egocentrism and the need for many students to control materials or dominate actions, the FOSS kit includes a lot of materials. To effectively manage students and materials, FOSS offers some suggestions.

Small-Group Centers

Many of the observations and investigations with materials for kindergarten are conducted with small groups at a learning center. Limit the number of students at the center to six to ten at one time. When possible, each student will have his or her own equipment to work with. In some cases, students will have to share materials and equipment and make observations together. Primary students are good at working together independently.

As one group at a time is working at the center on a FOSS activity, other students will be doing something else. Over the course of an hour or more, plan to rotate all students through the center, or allow the center to be a free-choice station.

Whole-Class Activities

Introducing and wrapping up the center activities require you to work for brief periods with the whole class. FOSS suggests for these introductions and wrap-ups that you gather the class at the rug or other location in the classroom where students can sit comfortably in a large group.

Guides for Adult Helpers

In the *Teacher Resources* binder, you will find duplication masters for center instructions sheets for some investigation parts. These sheets are intended as a quick reference for a family member or other adult who might be supervising the center. The sheets help that person keep the activity moving in a productive direction. The sheets can be laminated or slipped into a clear plastic sheet protector for durability.

When You Don't Have Adult Helpers

Some parts of investigations are designed for small groups, with an aide or a student's family member available to guide the activity and to encourage discussion and vocabulary development. We realize that there are many primary classrooms in which the teacher is the only adult present. Here are some ways to manage in that situation.

- Invite upper-elementary students to visit your class to help with the activities. They should be able to read the center instructions sheets and conduct the activities with students. Remind older students to be guides and to let primary students do the activities themselves.

- Introduce each part of the activity with the whole class. Set up the center as described in the *Investigations Guide*, but let students work at the center by themselves. Discussion may not be as rich, but most of the centers can be done independently by students once they have been introduced to the process. Be a 1-minute manager, checking on the center from time to time, offering a few words of advice or direction.

When Students Are Absent

When a student is absent for an activity, give him or her a chance to spend some time with the materials at a center. Another student might act as a peer tutor. Allow the student to bring home a *FOSS Science Resources* book to read with a family member.

SAFETY IN THE CLASSROOM AND OUTDOORS

Following the procedures described in each investigation will make for a very safe experience in the classroom. You should also review your district safety guidelines and make sure that everything you do is consistent with those guidelines. Two posters are included in the kit: *Science Safety* for classroom use and *Outdoor Safety* for outdoor activities.

Look for the safety icon in the Getting Ready and Guiding the Investigation sections that will alert you to safety considerations throughout the module.

Materials Safety Data Sheets (MSDS) for materials used in the FOSS Program can be found on FOSSweb. If you have questions regarding any MSDS, call Delta Education at 800-258-1302 (Monday–Friday, 8 a.m.–6 p.m. EST).

Science Safety in the Classroom

General classroom safety rules to share with students are listed here.

1. Listen carefully to your teacher's instructions. Follow all directions. Ask questions if you don't know what to do.

2. Tell your teacher if you have any allergies.

3. Never put any materials in your mouth. Do not taste anything unless your teacher tells you to do so.

4. Never smell any unknown material. If your teacher tells you to smell something, wave your hand over the material to bring the smell toward your nose.

5. Do not touch your face, mouth, ears, eyes, or nose while working with chemicals, plants, or animals.

6. Always protect your eyes. Wear safety goggles when necessary. Tell your teacher if you wear contact lenses.

7. Always wash your hands with soap and warm water after handling chemicals, plants, or animals.

8. Never mix any chemicals unless your teacher tells you to do so.

9. Report all spills, accidents, and injuries to your teacher.

10. Treat animals with respect, caution, and consideration.

11. Clean up your work space after each investigation.

12. Act responsibly during all science activities.

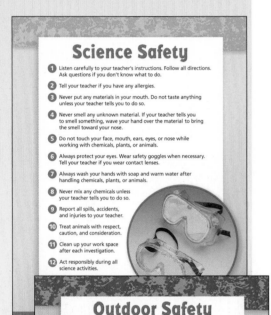

SCHEDULING THE MODULE

The Getting Ready section for each part of the investigation helps you to prepare. It provides information on scheduling the investigations and introduces the tools and techniques used in the investigation. The first item in the Getting Ready section gives an estimated amount of time the part should take. A general rule of thumb is to plan 10 minutes to introduce the investigation to the whole class, about 20–25 minutes at the center for each group, about 10 minutes to wrap up the activity with the whole class, and a few minutes to transition to the groups.

The investigations are numbered, and we suggest that they be conducted in that order, as the concepts build upon each other from investigation to investigation. Below is a list of the investigations and parts and the format of the investigation (whole class or center).

INVESTIGATION	PART	ORGANIZATION
1. Getting to Know Wood	1. Introduction to Wood	whole class
	2. Wood Hunt	whole class
	3. Wood and Water	center
	4. Sink the Pine and Plywood	center
2. Changing Wood	1. Sanding Wood	center or outdoors
	2. Sawdust and Shavings	center
	3. Making Sawdust Wood	center
	4. Making Sandwich Wood	center
3. Getting to Know Paper	1. Paper Hunt	whole class
	2. Using Paper	center
	3. Paper and Water	center
	4. Paper Recycling	center
	5. Papier-Mâché	center
4. Getting to Know Fabric	1. Feely Boxes and Fabric Hunt	center
	2. Taking Fabric Apart	center
	3. Water and Fabric	center
	4. Soiling and Washing Fabric	center or outdoors
	5. Graphing Fabric Uses	whole class
5. Earth Materials	1. Exploring Earth Materials	whole class
	2. Soil Painting	whole class
	3. Changes to Water	whole class
	4. Reuse and Recycle Resources	whole class
	5. Making Sculptures	center

FOSS K–8 SCOPE AND SEQUENCE

Grade	Physical Science	Earth Science	Life Science
6–8	Electronics Chemical Interactions Force and Motion	Planetary Science Earth History Weather and Water	Human Brain and Senses Populations and Ecosystems Diversity of Life
4–6	Mixtures and Solutions Motion, Force, and Models Energy and Electromagnetism	Weather on Earth Sun, Moon, and Planets Soils, Rocks, and Landforms	Living Systems Environments
3	Measuring Matter	Water	Structures of Life
1–2	Balance and Motion Solids and Liquids	Air and Weather Pebbles, Sand, and Silt	Insects and Plants Plants and Animals
K	Materials in Our World	Trees and Weather	Animals Two by Two

Materials

Contents

INTRODUCTION

The Materials in Our World kit contains

- *Teacher Toolkit: Materials in Our World*

    1 *Investigations Guide: Materials in Our World*

    1 *Teacher Resources: Materials in Our World*

    1 *FOSS Science Resources: Materials in Our World*

- *FOSS Science Resources: Materials in Our World*
  (1 big book and class set of student books)

- Equipment for 32 students

A new kit contains enough consumable items for at least two classroom uses before you need to resupply. Many of the FOSS early-childhood investigations take place at a science center for groups of six to ten students at a time. For whole-class activities, use a materials station for the class materials.

Individual photos of each piece of FOSS equipment are available online for printing. For updates to information on materials used in this module and access to the Materials Safety Data Sheets (MSDS), go to www.FOSSweb.com. Links to replacement-part lists and customer service are also available on FOSSweb.

▶ **NOTE**
Delta Education Customer Service can be reached at 1-800-258-1302.

KIT INVENTORY *List*

Drawer 1—print materials

Equipment Condition

1	*Teacher Toolkit: Materials in Our World* (1 *Investigations Guide*,
1	*Teacher Resources*, and 1 *FOSS Science Resources: Materials in Our World*)
32	*FOSS Science Resources: Materials in Our World*, student books*
1	*FOSS Science Resources: Materials in Our World*, big book

* The student books are shipped separately in one box of 32 soft cover books.

Drawer 1—permanent equipment

Equipment Condition

400	Paper clips, jumbo
70	Plates, paper ✪
5	Plywood pieces, 1/4" × 3/4" × 2.5"
2	Posters, FOSS Safety (*Science Safety* and *Outdoor Safety*)
1	Poster, *Cedar Tree*
1	Poster, *Linden Tree*
1	Poster, *Particleboard Production*
1	Poster, *Pine Tree*
1	Poster, *Plywood Production*
200	Rubber bands, #8
10	Screens
2	Self-stick notes, pads, 100/pad ✪
32	Spoons, metal
1	Tree round, aspen
36	Wood samples, basswood, 3/4" × 3/4" × 2.5"
36	Wood samples, cedar, 3/4" × 3/4" × 2.5"
36	Wood samples, particleboard, 3/4" × 3/4" × 2.5"
36	Wood samples, pine, 3/4" × 3/4" × 2.5"
36	Wood samples, plywood, 3/4" × 3/4" × 2.5"

Drawer 1—consumable equipment

5	Sandpaper, sheets, coarse, #50, 23 × 28 cm
2	Sawdust, bags
1000	Sticks, craft
200	Wood pieces, thin, 1/16" × 3/4" × 2.5"
36	Wood samples, basswood, for sanding
2	Wood shavings, bags

✪ These items might occasionally need replacement.

Drawer 2—permanent equipment

Qty	Item	Equipment Condition
6	Basins, clear plastic, 6 L (1.5 gal.)	
14	Bottles, clear plastic, with caps, 120 mL (4 oz.)	
8	Bottles, plastic squeeze, 60 mL (2 oz.)	
8	Brushes, scrub, small	
10	Containers, plastic, 1/4 L (8 oz.)	
32	Containers, plastic, 1/2 L (17 oz.)	
12	Containers, plastic, fluted, 1 L (34 oz.)	
25	Cups, plastic, 250 mL (9 oz.)	
36	Droppers, plastic	
10	Fabric squares, blue, 10 cm (4"), of each of these 10 kinds: burlap, corduroy, denim, fleece, knit, ripstop nylon, satin, seersucker, sparkle organza, terry cloth	
10	Lids for cups	
5	Loupes/magnifying lenses	
20	Magnets, doughnut-shaped	
1	Pebbles, large 15–30 mm diameter, bag, 2.3 kg/bag	
1	Pen, marking, permanent	
5	Sponges, large	
1	String, ball ✪	
1	Tape dispenser (for removable tape)	
1	Tape, masking, roll ✪	
1	Tape, removable, roll, 33 m/roll (1296"/roll) ✪	
5	Tape, transparent, rolls, 16.5 m/roll (650"/roll) ✪	
50	Zip bags, 1 L (1 qt.)	

Drawer 2—consumable equipment

Qty	Item	
75	Fabric squares, burlap, red, 4 cm (1.5")	
75	Fabric squares, muslin, 50%–100% cotton, 15 cm (6")	
75	Fabric squares, wool plaid, 4 cm (1.5")	
200	Index cards	
75	Paper samples, 10 cm square (4"), of each of these 6 kinds: chipboard, cardboard, corrugated, kraft, newsprint, tagboard	
1	Waxed paper, roll	

✪ These items might occasionally need replacement.

MATERIALS *Supplied by the Teacher*

Each part of each investigation has a Materials section that describes the materials required for that part. It lists materials needed for each student or group of students and for the class.

Be aware that you must supply some items. Each of these items is indicated with an asterisk (★) in the materials list for each part of the investigation. Here is a summary list of those items by investigation.

For most investigations
- Chart paper and marking pen
- Drawing utensils (crayons, pencils, colored pencils, marking pens)
- Glue sticks ★
- Glue, white ★
- Newspaper
- Paper towels
- 1 Pitcher or empty 2 L soft-drink bottle
- Science notebooks (composition books)
- 1 Scissors
- Water

For outdoor investigations
- Collecting bags for carrying materials
- Clipboards (optional)

Investigation 1: Getting to Know Wood
- 1 Turkey baster (optional)

Investigation 2: Changing Wood
- Container, large zip bag, or jar
- Cornstarch, 1 or 2 boxes
- 1 Plane (optional)
- Safety goggles
- Saucepan
- 1 Saw (optional)
- Scratch paper
- 1 Spoon, long handled
- 2 Spoons, plastic
- 32 Sticks, about 20 × 1 cm (8" × 0.5")

Investigation 3: Getting to Know Paper

1   Clothesline and clothespins (optional)

•   Construction paper, white

•   Facial tissue, white

•   Flour, 3/4–1 L (3–4 cups)

•   Newsprint (plain) or brown paper towels (from school)

1   Paper, white, sheet, 22 × 28 cm (8.5" × 11")

1   Screwdriver or mat knife

1   Spoon, large, with long handle

1   Tissue paper, roll, single ply

Investigation 4: Getting to Know Fabric

1   Box cutter

5   Cardboard boxes or brown-paper grocery bags

•   Clothesline and clothespins (optional)

2   Contact paper (optional)

•   Liquid laundry detergent or dish soap, 3/4 L (3 cups)

•   Staining materials (mustard, ketchup, tempera paint, grape juice)

1   Stapler (optional)

Investigation 5: Earth Materials

1   Balance (optional)

•   Fabric scraps, small

•   Garden trowel (optional)

•   Ice chest or cooler (optional)

•   Mat-board scraps (optional)

•   Paper plates or white construction paper, 28 × 45 cm (11" × 18")

•   Paper scraps

•   Materials for recycling center (empty plastic bottles, steel cans, aluminum cans, newspapers, cardboard boxes, scrap wood, fabric scraps, clean rags)

•   Soil

•   Towel, cloth

•   Wood scraps, small

PREPARING *a New Kit*

If you choose to prepare the materials all at once with a group of volunteers, you can use the following guidelines for organization.

1. **Prepare the center instruction sheets**
   Each investigation part that involves a group of students at a center has a center instructions sheet written for a parent or other adult helper working with students. Each sheet summarizes and abbreviates the information provided to the teacher in the guide (and is not a replacement for the teacher-guide instructions). Use the teacher masters to make copies of the center instructions and either laminate the sheets or put them in clear-plastic sheet protectors. Take time to orient your adult volunteers or aides to the overall purposes of the activities, and encourage them to facilitate but not direct student learning at the center. Below are the numbers for the center instruction sheets.

   No. 6   Center Instructions—Wood and Water

   No. 7   Center Instructions—Sink the Pine and Plywood

   No. 8   Center Instructions—Sinking Investigation

   No. 10 Center Instructions—Sanding Wood

   No. 11 Center Instructions—Sawdust and Shavings

   No. 12 Center Instructions—Making Sawdust Wood

   No. 13 Center Instructions—Making Sandwich Wood

   No. 16 Center Instructions—Using Paper A

   No. 17 Center Instructions—Using Paper B

   No. 18 Center Instructions—Paper and Water

   No. 19 Center Instructions—Paper Recycling

   No. 20 Center Instructions—Papier-Mâché

   No. 29 Center Instructions—Feely Boxes and Fabric Hunt

   No. 31 Center Instructions—Taking Fabric Apart

   No. 32 Center Instructions—Water and Fabric

   No. 33 Center Instructions—Soiling and Washing Fabric A

   No. 34 Center Instructions—Soiling and Washing Fabric B

   No. 37 Center Instructions—Making Sculptures

2. Prepare wood samples

Six bags of wood samples are included in the kit. Five bags of samples (basswood, particleboard, pine, plywood, and cedar) are permanent materials and should not be sanded. A second bag of basswood is consumable. It should be labeled "FOR SANDING."

The wood samples are all the same size and shape so that students will focus on the properties of the different kinds of wood. Most of the samples are easily recognized, but you may have trouble distinguishing between pine and basswood. If you are the first to use the kit, use a permanent marker to put a tiny dot on one end of each basswood sample for positive identification later.

PREPARING *the Kit for Your Classroom*

Some preparation is required each time you use the kit. Doing these things before beginning the module will make daily setup quicker and easier.

1. **Inventory materials**

   Before using a kit, conduct a quick inventory of all items in the kit. You can use the list provided in this chapter to keep track of any items that are missing or in need of replacement. Information on ordering replacement items can be found at the end of this chapter. The kit contains enough consumables for at least two classes of 32 students.

2. **Check basswood samples**

   Two bags of basswood samples are in the kit. One is permanent equipment and should not be sanded or modified in any way. A second bag of basswood is consumable, although the samples can be reused several times. Students will change the shape of these by sanding them in Investigation 2. Check the label on the bag to be certain you are using the basswood that is marked "FOR SANDING" when you begin the investigation.

3. **Prepare consumable paper samples**

   Most of the paper samples in the kit are precut into 10-centimeter (cm) (4") squares. There are 75 samples of each kind of paper in the kit—enough for at least two classroom uses. A roll of waxed paper is provided in the kit, and you will need to cut it into 10 cm squares. You will also need to provide construction-paper samples, paper-towel samples, and facial-tissue samples. Cut 50 squares of each paper. This should give you enough samples for the class, plus several extras if needed. All materials can be purchased from local sources, and replacement packages are available from Delta Education.

4. **Check permanent fabric samples**

   All the fabric samples in the kit come precut; samples of each of the ten kinds of blue fabric come in their own zip bags. You will assemble sets of five or ten fabrics, depending on the activity. The Getting Ready sections will tell you how to group the samples for each part. The edges of the fabric may become a bit frayed from use, but if you caution students to handle them carefully, the samples will not need to be replaced very often. The ten blue fabrics are considered to be permanent equipment and should not be cut up for fabric projects.

5. **Check consumable fabric samples**

   The small burlap and wool-plaid squares and the larger muslin squares provided in the kit will be consumed in the course of the activities. These fabrics should be resupplied after two classes have used the kit. All these consumable materials can be purchased from local sources, but replacement packages are available from Delta Education.

6. **Care and reuse of materials**

   The items in the kit have been selected for their ease of use and durability. Make sure that items are clean and dry before putting them back in the kit. Small items should be inventoried (a good job for students under your supervision) and put in zip bags for storage.

7. **Identify containers**

   A variety of containers are in the kit. The first time they are used, they are described and illustrated for easy identification. Metric capacities are used to distinguish the containers. We also include the English equivalents for the first time they are described in an investigation. Be sure to use the correct container for each investigation and, if you have questions, check the photo equipment cards.

8. **Plan for drying time**

   In a number of activities, you will need to allow time and space for wood, paper, fabric, or rocks to dry. Be sure to read ahead and plan for this time.

   > Investigation 1, Part 3: Wood and Water
   >
   > Investigation 1, Part 4: Sink the Pine and Plywood
   >
   > Investigation 1, Part 5: Sinking Investigation
   >
   > Investigation 2, Part 2: Sawdust and Shavings
   >
   > Investigation 3, Part 3: Paper and Water
   >
   > Investigation 4, Part 3: Water and Fabric
   >
   > Investigation 4, Part 4: Soiling and Washing Fabric
   >
   > Investigation 5, Part 1: Exploring Earth Materials

Materials in Our World Module

In other activities, you will need to allow time and space for projects to dry before they can go home.

Investigation 2, Part 3: Making Sawdust Wood

Investigation 2, Part 4: Making Sandwich Wood

Investigation 3, Part 4: Paper Recycling

Investigation 3, Part 5: Papier-Mâché

Investigation 5, Part 2: Soil Painting

Investigation 5, Part 5: Making Sculptures

9. **Gather books from library**

Check your local library for the books related to this module. These books are recommended in the Interdisciplinary Extensions section of each investigation.
- *Math in Motion: Origami in the Classroom* by Barbara Pearl (Inv. 3)
- *Pezzettino* by Leo Lionni (Inv. 3)
- *Red Leaf, Yellow Leaf* by Lois Ehlert (Inv. 3)
- *Biggest, Strongest, Fastest* by Steve Jenkins (Inv. 3)
- *Caps for Sale* by Esphyr Slobodkina (Inv. 4)
- *No Roses for Harry!* by Gene Zion (Inv. 4)

10. **Plan for student creations**
Students produce several projects. Help students keep track of all the projects so that they have complete collections to take home.

11. **Plan for science notebooks**
See Getting Ready for Investigation 1, Part 1, for ways to organize the science notebooks for this module.

12. **Plan for the word wall and pocket chart**
As the module progresses, you will add new vocabulary words to a word wall or pocket chart and model writing and responding to focus questions. Plan how you will do this in your classroom. There are suggestions in Getting Ready for Investigation 1.

13. **Plan for focus-question charts**
Each part of each investigation has a focus question that students are asked before and after the activity session. You'll find these questions on teacher masters 2–4, *Focus Questions*. Students will glue each focus question on a page in their science notebooks and respond to it with words or drawings. At the beginning of the module, you will need to scaffold the use of notebooks. Use a chart to model how to respond to the focus question in writing or drawings. See Getting Ready for Investigation 1, Part 1, for suggestions on how to do this in your classroom.

TEACHING NOTE

Refer to the teacher resources on FOSSweb for a list of appropriate trade books that relate to this module.

14. Plan for letter home and home/school connections

Teacher master 1, *Letter to Family*, is a letter you can use to inform families about this module. The letter states the goals of the module and suggests some home experiences that can contribute to students' learning. Space is left at the top so you can copy the letter onto your school letterhead.

There is a home/school connection for most investigations. Check the last page of each investigation for details, and plan when to make copies and send them home with students.

15. Consider safety issues indoors and outdoors

Early-childhood students should be allowed to demonstrate that they can act responsibly with materials, but they must be given guidelines for safe and appropriate use of materials. Work with students to develop those guidelines so they can participate in making behavioral rules and understand the rationale for the rules. Emphasize that materials do not go in mouths, ears, noses, or eyes. Encourage responsible actions toward other students.

Two safety posters are included in the kit to post in the room—*Science Safety* and *Outdoor Safety*. The Getting Ready for Investigation 1, Part 1, will offer suggestions for this discussion. Also be aware of any allergies that students in your class might have. Students with latex allergies should not handle the rubber bands.

Three investigation parts require special consideration. In Investigation 2, Parts 1, 2, and 3, students are exposed to sawdust and wood shavings. Find out which students have asthmatic conditions, and be sure to accommodate their health needs. Check your district guidelines for use of safety goggles while sanding wood.

16. Check FOSSweb for resources

Go to FOSSweb and review the print and digital resources available for this module.

> ▶ **NOTE**
> The *Letter to Family* and *Home/School Connections* are also available electronically on FOSSweb.

CARE, *Reuse, and Recycling*

When you finish teaching the module, inventory the kit carefully. Note the items that were used up, lost, or broken, and immediately arrange to replace the items. Use a photocopy of the Kit Inventory List in this chapter, and put your marks in the "Equipment Condition" column. Refill packages and replacement parts are available for FOSS by calling Delta Education at 1–800–258–1302 or by using the online replacement-part catalog (www.DeltaEducation.com).

Standard refill packages of consumable items are available from Delta Education. A refill package for a module includes sufficient quantities of all consumable materials (except those provided by the teacher) to use the kit with two classes of 32 students.

Here are a few tips on storing the equipment after use.

- Make sure items are clean and dry before storing them.
- Make sure the posters and print materials are flat on the bottom of the box.
- Return wood samples, paper samples, and fabric samples to the appropriately labeled bags.
- Inventory and bag up the paper and fabric scraps that can be used for the next class. Extra zip bags are included for this purpose.
- Carefully seal the sawdust and wood-shavings bags.
- Return paper clips and small rubber bands to their bags.
- Label the dispenser of removable tape so it is not confused with regular tape.
- Inventory and bag the small items.

The items in the kit have been selected for their ease of use and durability. Make sure that items are clean and dry before putting them back in the kit. Small items should be inventoried (a good job for students under your supervision) and put into zip bags for storage. Any items that are no longer useful for science should be properly recycled.

Investigation 1: Getting to Know Wood

PURPOSE

Content

- Wood can be described in terms of its properties.

- Different kinds of wood come from different kinds of trees. Trees are natural resources. Some kinds of wood are processed and made by people.

- Wood floats in water but can be made to sink.

- Wood absorbs water.

Scientific Practices

- Observe and compare physical properties of different kinds of wood samples, using the senses.

- Compare properties of different kinds of wood found in the classroom and outdoor environments.

- Observe how wood interacts with water.

- Describe relative position, using one reference.

- Communicate observations made about different kinds of wood, orally and through drawings.

	Investigation Summary	Time	Focus Question
PART 1	**Introduction to Wood** Students become familiar with different kinds and forms of wood found in their home and school environments. Students compare and describe five uniform samples of different kinds of wood, learn the names, and observe how the woods are alike and how they are different.	**Introduction** 10 minutes **Class/Center** 20 minutes **Notebook** 15 minutes	Where does wood come from?
PART 2	**Wood Hunt** Students go on a wood hunt and label objects in the classroom that are made of wood. They also go outside to find wood.	**Indoor Hunt** 20 minutes **Outdoor Hunt** 15 minutes **Reading** 15 minutes	What is made of wood?
PART 3	**Wood and Water** Students observe how wood and water interact, first by putting drops of water on the wood, then by putting the wood in basins of water.	**Introduction** 10 minutes **Center** 15 minutes **Outdoors** 15 minutes **Notebook** 15 minutes	What happens when wood gets wet?
PART 4	**Sink the Pine and Plywood** Students find ways to sink two of the floating wood samples by attaching paper clips to the wood with rubber bands. Students discover how easy it is to sink the plywood compared to the pine sample. They use relative-position words to describe the location of the wood.	**Introduction** 10 minutes **Center** 30 minutes **Notebook** 15 minutes	How can you sink wood?
PART 5	**Sinking Investigation (Optional)** Students refine their techniques of using paper clips and rubber bands to sink wood samples, and they test two kinds of wood. Students make a bar graph of their results, lining up and comparing the number of clips used.	**Introduction** 10 minutes **Center** 20 minutes **Notebook** 15 minutes	How many paper clips does it take to sink wood?

Content	Writing/Reading	Assessment
• Wood has many observable properties. • Wood is a resource that comes from different kinds of trees. • Some kinds of woods are processed and transformed by people.	**Science Notebook Entry** Draw or write words to answer the focus question.	**Embedded Assessment** Teacher observation
• Wood has many observable properties. • Wood is used for many everyday things.	**Science Notebook Entry** Draw or write words to answer the focus question. **Science Resources Book** "The Story of a Chair"	**Embedded Assessment** Teacher observation
• Wood has many observable properties. • Wood floats in water. • Wood absorbs water.	**Science Notebook Entry** Draw or write words to answer the focus question.	**Embedded Assessment** Teacher observation
• Wood has many observable properties. • Wood floats in water. Some kinds of wood sink more easily than others.	**Science Notebook Entry** Draw or write words to answer the focus question.	**Embedded Assessment** Teacher observation
• Wood floats in water. Some kinds of wood sink more easily than others.	**Science Notebook Entry** Draw or write words to answer the focus question.	**Embedded Assessment** Teacher observation

BACKGROUND *for the Teacher*

Where Does Wood Come From?

Wood is the name given to tree parts used for fuel or building construction. Wood is an important renewable resource used in countless ways around the world.

Cellulose is a complex carbohydrate closely related to starch. Like starch, cellulose is created by plants. Most of the tough, stiff parts of plants are made of cellulose fibers. The cellulose fibers are surrounded by lignin, an organic compound that acts like glue holding the fibers together, adding to the strength of wood. The thick layers of reinforced cellulose laid down in the cell walls of trees and shrubs give trunks, branches, and twigs their characteristic rigid structure.

Cotton, seed coverings, stems, and leaves contain large quantities of cellulose. Cellulose resists digestion by most animals, including humans, so it is nearly useless as a food source. However, multitudes of microorganisms (bacteria and fungi) can biochemically break down cellulose into simple chemicals that can be used for food, thus providing vitally important nutrient recycling in the environment.

To most of us, the notion of wood produces a generic image of a lightweight, tan material with grain lines running through it. However, wood comes in a wide variety of natural colors, densities, and patterns. While most wood **floats** in water, ebony and mesquite are denser than water and **sink** to the bottom when pitched into a tub or pool. Wood can be tan, like oak, fir, and ash, but it can also show amazing color variation: black, red, pink, purple, orange, white, green, and yellow. Many kinds of wood display an attractive pattern, called **grain**, that results from the characteristic dark and light concentric growth lines laid down in the trunks of trees. Some trees have imperceptible grain, such as balsa and ebony. Others have dramatic patterns of circles, spots, zigzags, and random designs.

> *"Wood is the name given to tree parts used for fuel or building construction. Wood is an important renewable resource used in countless ways around the world."*

This investigation uses three kinds of natural wood: cedar, pine, and basswood. Cedar and pine are known as softwoods (also known as gymnosperm **trees**) because they come from coniferous trees—trees with needlelike or scalelike leaves. **Pine** is a common, versatile wood that comes from any one of a family of related trees. It is used for construction and furniture. Its desirable **properties** include availability, lightweight, strength, and ease of working. **Cedar** is naturally weather- and pest-resistant, so it is an excellent choice for outdoor applications such as decking and fencing. **Basswood** is a hardwood (angiosperm tree) that is milled from broadleaf linden trees that grow in the eastern forests. Ironically, this hardwood is a soft, almost grainless wood used by carvers and model builders. All three of these woods are native to North America.

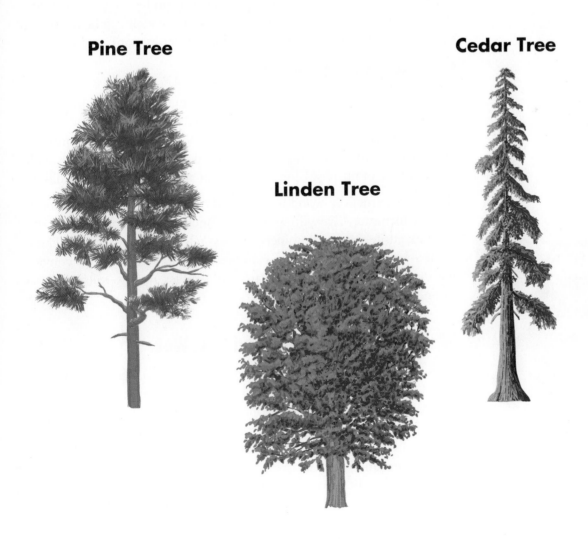

Pine Tree

Linden Tree

Cedar Tree

Some kinds of wood are manufactured products. **Plywood** is made from wood and glue and has become the worldwide standard for a large-surface, flat piece of wood. Because of the layers involved in producing plywood, children often call it sandwich wood.

Plywood is made in an interesting way. Imagine an ear of corn on the cob with a couple of those little handles you poke into the ends of the cob. Now imagine a large log suspended the same way and clamped into a machine that will rotate the log like a rolling pin. A huge knife the length of the log is brought up to the side of the rotating log, and a thin sheet of wood comes peeling off like paper towels coming off the roll. The process continues until the whole log has been transformed into flat sheets of wood a few millimeters thick. The thin sheets of wood are glued together, alternating the direction of the grain, to make a large, exceptionally strong sheet of wood.

Waste wood in the form of chips and sawdust can be mixed with glue and made into a product called **particleboard**, which can be formed into sheets. Students call it sawdust wood for obvious reasons.

What Is Made of Wood?

A great technological breakthrough was the milling of wood to make lumber. Lumber is wood cut to uniform sizes for the purpose of construction. We tend to use the words *wood* and *lumber* synonymously, but wood is a **material** that can take a multitude of forms and be applied to a universe of purposes; lumber is boards and beams used for building.

Besides construction, wood is used for lots of consumer products, most notably furniture. Everything from perfunctory tables and stools to the most exquisitely crafted cabinets and mantels are made of wood. The endless variety of small household objects brings everyone into contact with wood every day: kitchenware, tool handles, pencils, rulers, toys, and so on.

What Happens When Wood Gets Wet?

Wood's two primary functions are to give the tree structure and to convey liquids, and the substances dissolved in them, throughout the plant's structure. If you train a microscope on a cross section of a tree trunk or branch, you will see thousands of circular cells, which are the openings of long tubes. Some of these vessels (xylem) carry water and minerals from the ground to the leaves; the others (phloem) carry sugars and other nutrients from the leaves to the rest of the plant. When the vessels cease to carry fluids, the tissue is filled with empty pipelines. That accounts for the porosity of wood.

When you put a drop of water on wood, several things might occur depending on the kind of wood, how it is prepared, and where the water is applied. Softwoods tend to have lots of access points for water to enter, so it is readily **absorbed**. A drop placed parallel to the grain will spread out quickly, moistening an elongated area as it does so. The same drop placed on the end of a piece of wood may not spread laterally, but will disappear quickly into the wood. Denser woods, particularly those containing natural waxes or resins, may resist absorption for a time, allowing the water to **bead** up and persist. Manufactured wood products, such as particleboard, are high in adhesive compounds that may render the wood impervious for quite a while.

Wood is resilient. After being doused, wood will dry, and for a substantially long time, it will maintain its integrity and strength. It is not the water that causes wood to rot, but rather the microorganisms that move in, eventually eating themselves out of house and home.

How Can You Sink Wood?

There are essentially two ways to sink a piece of wood. You can overcome its buoyancy by attaching dense objects to the sample, or you can wait for nature to do the deed. The materials from which wood is made—cellulose and lignin—are more dense than water. It's the air inside the abandoned xylem and phloem vessels that makes a sample of wood less dense than water. When wood is tossed into water, the water slowly but surely invades the air spaces, displacing the air. When water has infiltrated the porous spaces sufficiently to tip the density to 1.0 grams per cubic centimeter, wood is considered waterlogged, and gravity will pull the wood to the bottom of the body of water.

TEACHING CHILDREN *about Wood*

Wood is so common that its importance in everyday life can easily be overlooked. One of the two reciprocating roles of science for early-childhood students is to provide experiences that transform familiar objects, materials, and events into engaging subjects for extended inquiry. Wood provides just such an opportunity.

Early-childhood students know wood when they see it—at least most kinds of wood. They know it as the material used to make tables, doors, blocks, broom handles, floors, and a host of other everyday things. By observing wood samples closely, students begin to understand that there are similarities among all kinds of woods, and that, at the same time, unique properties make each kind of wood identifiable. As with all materials, different kinds of wood behave in particular ways when they are allowed to interact with other materials.

Early-childhood students love to put just about anything into water. Wood is wonderful for water investigations, and students will want to spend plenty of time watching what wood does in water. Wood changes color when it gets wet, and often patterns of grain, particle, or layer are revealed much more clearly. The little blocks of wood used in this activity will become swimmers, divers, boats, towers, spaceships, and countless other objects, all of which end up in the water as a result of some series of imaginative events. In the process of exploring their scenarios, students will observe floating, splashing, bobbing, and rolling as the wood pieces are repeatedly put into and taken out of the water. Students are also fascinated by the fact that two or more pieces of wet wood will cling to one another as if they were sticky or magnetic.

An important part of investigating floating is attempting to defeat wood's ability to float. It is through such systematic planning and problem solving that students come to a higher level of understanding of the materials and the principles governing the behavior of those materials. When students are challenged to sink pieces of wood, they find out more about wood and start to develop an intuitive understanding of buoyancy. (The complete scientific understanding of buoyancy will not be accessible to students for a number of years.)

Vocabulary is most effectively developed when it is introduced in a naturalistic way. By hearing a new word in an unambiguous context, students advance their language skills rapidly. During the hands-on activities, a number of new and familiar vocabulary words will be used.

New Word
Say it · See it · Hear it · Write it

Above
Absorb
Basswood
Bead
Below
Cedar
Communicate
Compare
Different
Fewer
Float
Grain
Graph
Less
Material
More
Observe
Particleboard
Pine
Plywood
Property
Rough
Same
Senses
Sink
Smooth
Soak
Spread
Test
Texture
Tree
Wood

The most important outcome of this activity might be the opening of young students' eyes to the existence of wood everywhere. Wood is a material that plays a central role in virtually every culture around the world because of its availability and the ease with which it can be transformed into objects of art, commerce, defense, transportation, and housing. As a result of direct personal experience with wood, students will have an elevated awareness of this important material.

The ideas presented in the **Materials in Our World Module** are useful to young scientists as they begin the long, intriguing process of imposing order and reason on the daunting complexity of the natural world. The concepts build on one another. They have coherence and direction. Students will benefit greatly when the concepts are developed in the suggested sequence. We refer to this sequence as the conceptual flow of the module.

The **conceptual flow** for this first investigation starts with **wood**. In Part 1, wood is identified as a **kind of material**. Students **observe** five different samples of wood, finding that different kinds of wood have different **properties**: **color**, **pattern**, **smell**, **texture**. The properties can be used to describe or identify wood. Students think about trees as the **source of wood** and learn that **natural** wood comes from trees, and that manufactured wood is **made by people** in factories.

In Part 2, students search their classroom and the outdoor environment to see where wood is **used** and what objects and structures are made from it.

In Part 3, students apply drops of water to wood to find out how **water interacts with wood**. They find that the water may **bead up**, **spread out**, or **soak into** the wood.

Students plunge wood samples into water in Part 4 in order to investigate how **wood interacts with water**. Students discover that wood **floats**, but with ingenuity and perseverance they can make it **sink**. They find that it takes more paper clips to sink some kinds of wood than others.

Part 5 is a systematic investigation of wood sinking. Students establish a process and gather data. The data are then displayed in a graph. This may be most effective if it is postponed until students are a little older (perhaps second semester when they are five-and-a-half-year-olds).

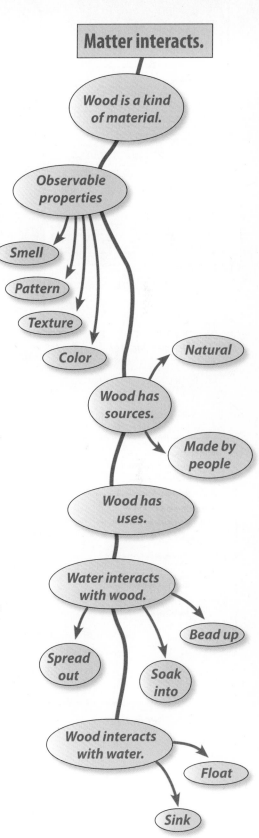

LETTER TO FAMILY

.... Science News ....

Dear Family,

Our class is beginning a study of materials. We will be studying the properties of wood, paper, fabric, soil, and rock and how different materials are alike and how they are different. We'll investigate how wood and paper can be processed into products. We'll sand wood, make simulated plywood and particleboard, recycle paper, and make paper-mâché bowls. We will study how fabrics are made and discover how they interact with water. For several weeks, we will be materials scientists.

During our materials study, we will focus on the reuse and recycling of materials to conserve natural resources. We will be setting up a recycling center in our classroom. You can enrich this experience by having your child participate in the recycling of paper, metal, glass, and plastic at home.

We can also use help gathering wood scraps and interesting and colorful paper and fabric discards for making our final projects. Please send wood, paper, and fabric scraps by _____ (date).

After we do the various activities in class, your child may ask you to help him or her do things at home, such as temporarily label objects with "paper" or "wood"; waterlog a stick; take boxes apart; or make collages, envelopes, drinking cups, or paper hats. You can help by making a few simple materials available and letting your child be creative.

Sincerely,

.... Science News ....

Dear Family,

Our class is beginning a study of materials. We will be studying the properties of wood, paper, fabric, soil, and rock and how different materials are alike and how they are different. We'll investigate how wood and paper can be processed into products. We'll sand wood, make simulated plywood and particleboard, recycle paper, and make paper-mâché bowls. We will study how fabrics are made and discover how they interact with water. For several weeks, we will be materials scientists.

During our materials study, we will focus on the reuse and recycling of materials to conserve natural resources. We will be setting up a recycling center in our classroom. You can enrich this experience by having your child participate in the recycling of paper, metal, glass, and plastic at home.

We can also use help gathering wood scraps and interesting and colorful paper and fabric discards for making our final projects. Please send wood, paper, and fabric scraps by _____ (date).

After we do the various activities in class, your child may ask you to help him or her do things at home, such as temporarily label objects with "paper" or "wood"; waterlog a stick; take boxes apart; or make collages, envelopes, drinking cups, or paper hats. You can help by making a few simple materials available and letting your child be creative.

Sincerely,

No. 1—Teacher Master

FOCUS QUESTIONS A

Inv. 1, Part 1: **Where does wood come from?**

Inv. 1, Part 2: **What is made of wood?**

Inv. 1, Part 3: **What happens when wood gets wet?**

Inv. 1, Part 4: **How can you sink wood?**

Inv. 1, Part 5: **How many paper clips does it take to sink wood?**

Inv. 2, Part 1: **How can you change the shape of wood?**

Inv. 2, Part 2: **How are sawdust and shavings the same? How are they different?**

Inv. 2, Part 3: **How is particleboard made?**

Inv. 2, Part 4: **How is plywood made?**

No. 2—Teacher Master

MATERIALS *for*
Part 1: *Introduction to Wood*

For each student

  1 *Letter to Family*

  1 Science notebook (See Step 7 of Getting Ready.) ★

For the class

  5 Basins

 32 Basswood samples

 32 Particleboard samples

 32 Pine samples

 32 Plywood samples

 32 Cedar samples

  1 Poster, *Linden Tree*

  1 Poster, *Particleboard Production*

  1 Poster, *Pine Tree*

  1 Poster, *Plywood Production*

  1 Poster, *Cedar Tree*

  1 Tree round, aspen

  • Glue sticks ★

  • Chart paper ★

  2 Marking pens, different colors (for teacher) ★

  • Crayons and marking pens for students ★

  • *Science Safety* and *Outdoor Safety* posters

❏ 1 Teacher master 1, *Letter to Family*

❏ 1 Teacher master 2, *Focus Questions A* (See Step 7 of Getting Ready.)

For assessment

❏ • *Assessment Checklist*

★ Supplied by the teacher.   ❏ Use the duplication master to make copies.

FOSS

GETTING READY *for*
Part 1: *Introduction to Wood*

1. **Schedule the investigation**

   Part 1 is a whole-class activity, but it also works well with groups of six to ten students at a center. Plan on 20 minutes for working with wood and 15 minutes for recording in notebooks.

2. **Preview Part 1**

   Students become familiar with different kinds and forms of wood found in their home and school environments. Students compare and describe five uniform samples of different kinds of wood, learn the names, and observe how the woods are alike and how they are different. The focus question is **Where does wood come from?**

3. **Prepare wood samples for distribution**

   There are samples of five kinds of wood in the kit. You will need one sample of each kind of wood for each student. Put all the samples of each kind of wood in their own clear plastic basins. To start the activity, each student will walk by the five basins and pick up one piece of each kind of wood. If you set the basins on a small table, students can walk on both sides to pick up pieces, making the process go faster.

TEACHING NOTE

The Getting Ready section for Part 1 of Investigation 1 is longer than the corresponding section for the other parts. Several of the numbered steps appear only here, but may apply to other parts as well, such as planning a word wall (Step 6), preparing student notebooks (Step 7), photocopying duplication masters (Step 8), and planning for safety (Step 10).

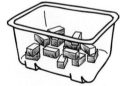    

4. **Become familiar with the wood samples**

   Study the five different samples of wood so you become familiar with the similarities and differences. Pine and basswood are very similar, but pine has the familiar smell of pitch. If you are the first to use the kit, use a permanent marker to put an inconspicuous dot at one end of each basswood sample for positive identification later. Keep the dots small, and don't point them out to students.

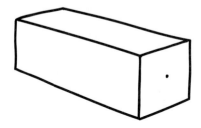

5. Check posters

Three tree posters and two wood-production posters are included in the kit. Review them in preparation for the investigation.

6. Plan to use a word wall or pocket chart

As the module progresses, you will add new vocabulary words to cards or sentence strips for use in a pocket chart or on a word-wall chart for posting on a wall or an easel. You will also use a focus chart to write the focus question for that day and to model responses. The focus chart can serve as a page in the class notebook to model making notebook entries. For additional information, see the Science-Centered Language Development chapter.

FOCUS CHART

Where does wood come from?

Wood comes from trees.

Pine comes from a _____ .

Plywood is made by _____ .

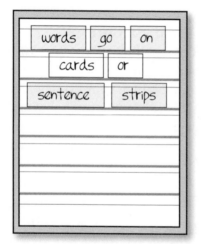

WORD WALL

words
words
and more words

7. Plan for student notebooks

Students will keep a record of their science investigations in their own science notebooks. Students will record observations and responses to focus questions. This record will be a useful reference document for students and a revealing testament for adults of each student's learning progress.

We recommend that students use bound composition books for their science notebooks. You can make individual science notebooks by stapling 12–15 blank pages together with a cover. This ensures that student work is maintained as an organized, sequential record of student learning.

If you are already using an alternative method of organization with your students, such as a sheath of folded and stapled pages, your method can take the place of the bound composition book.

▶ **NOTE**

See the *Science Notebooks in K–2* chapter for details on setting up and using notebooks.

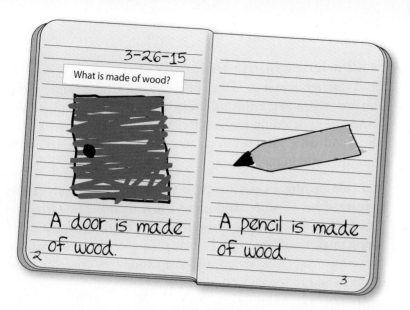

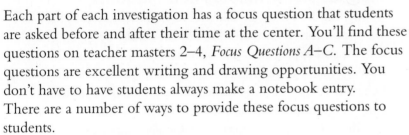

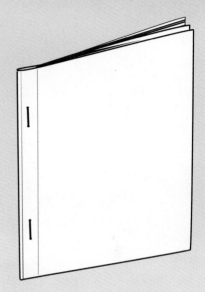

Each part of each investigation has a focus question that students are asked before and after their time at the center. You'll find these questions on teacher masters 2–4, *Focus Questions A–C.* The focus questions are excellent writing and drawing opportunities. You don't have to have students always make a notebook entry. There are a number of ways to provide these focus questions to students.

- Photocopy the focus question sheets, cut the focus questions apart, and have them ready for each student to glue into his or her notebook at the appropriate time in each part.

- Prepare sheets with one focus question on each sheet and room for students to draw and write. You could also include writing frames on each sheet. Writing frames are good ways to model notebook entries. Photocopy these sheets and assemble them into a science notebook for each student.

8. **Photocopy duplication masters**
Teacher masters serve various functions—letter to family, home/school connections, center instructions, and focus questions. These can be duplicated, cut apart, and glued into student notebooks. A master that requires duplication is flagged with this icon ❑ in the materials list for each part.

9. **Send a letter home to family**

   Read teacher master 1, *Letter to Family*, and send it home with students as you begin this investigation. The letter explains what students will be doing with materials such as wood, paper, and fabric, and what parents can do to extend the experiences at home with their children.

   As a culminating activity for the entire module, students create sculptures out of paper, wood, and fabric scraps. The *Letter to Family* asks families to send contributions of interesting paper scraps, small wood scraps, and fabric scraps for use in Investigation 5. Plan for storing these scraps, and keep your eyes open for additional scraps to add to the collection.

10. **Plan for safety indoors and outdoors**

    Young children must be allowed to demonstrate that they can act responsibly with materials, but they must be given guidelines for safe and appropriate use of materials. Work with students to develop those guidelines so that students participate in making behavior rules and understand the rationale for the rules. Encourage responsible actions toward other students. Display and discuss the *Science Safety* and *Outdoor Safety* posters in class.

    Look for the safety icon in the Getting Ready sections, which will alert you to safety concerns throughout the module. Be aware of any allergies your students have, including latex allergies. (Rubber bands are made of latex.)

11. **Plan for working with English learners**

    At important junctures in an investigation, you'll see sidebar notes titled "EL Note." These notes suggest additional strategies for enhancing access to the science concepts for English learners. Refer to the Science-Centered Language Development chapter for resources and examples to use when working with science vocabulary, writing, oral discourse, and readings.

    Each time new science vocabulary is introduced, you'll see the new-word icon in the sidebar. This icon lets you know not only that you'll be introducing important vocabulary, but also that you might want to plan on spending more time with those students who need extra help with the vocabulary.

12. **Assess progress throughout the module**

    Assessment opportunities are embedded throughout the module to help you look closely at students' progress. In kindergarten, those assessment opportunities involve teacher observations of students' actions with materials and verbal interactions with you

Say it
Write it
New Word
See it
Hear it

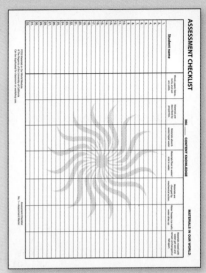

No. 1—Assessment Master

and other students, focusing on the materials being studied. Read through the Assessment chapter for a description of the assessment opportunities.

You will find the two-page *Assessment Checklist* in the Assessment Masters. The first page lists the science-content objectives that students will encounter throughout the module and described in each Getting Ready section. You can assess each of these objectives several times during the course of the module. There is also a space on the checklist for writing your observations for each student. The second page of the *Assessment Checklist* is for recording students' ability to engage in scientific practices appropriate for their age.

13. **Plan assessment for Part 1**

There are six objectives that can be assessed at any time during any part of this investigation.

What to Look For

- *Students ask questions.*
- *Students plan and conduct simple investigations.*
- *Students use senses to observe materials and objects.*
- *Students record and organize observations.*
- *Students communicate observations orally and in their notebooks with words and drawings.*
- *Students incorporate new vocabulary.*

Here are specific content objectives to observe in this part.

- *Wood has observable properties and can be described by those properties.*
- *Wood is a resource that comes from trees.*

Here are the specific scientific practices to observe in this part.

- *Students compare properties of wood (organize observations).*
- *Students sort wood by properties (organize observations).*

Make copies of the *Assessment Checklist*, attach them to a clipboard, and carry them with you when students are engaged in the investigations. Record your observations as you interact with students, or take a few minutes after class to reflect on the lesson. Because there are several opportunities for you to assess students on each objective, we suggest that you focus on six to ten students during each session rather than trying to assess the whole class at one time.

GUIDING *the Investigation*
Part 1: *Introduction to Wood*

1. **Introduce** *observe*

   Call students to the rug. If this is their first introduction to making observations of objects, introduce the word *observe* with a mini-lesson. One way to do this is to have students respond chorally. Below is an example of this mini-lesson.

   ➤ *Today we are going to learn a very important science word. Say "**observe**."* (Write the word on the word wall or place it in the pocket chart.)

   S: Observe.

   ➤ *Let's clap the parts (syllables) of the word.*

   S: Ob (clap) serve (clap).

   ➤ *Again.*

   S: Ob (clap) serve (clap).

   ➤ *How many parts (syllables) are in the word* observe?

   S: Two.

   ➤ *Observe means to look at something carefully. What does observe mean?*

   S: To look at something carefully.

   ➤ *When I look at something carefully, I _____ .*

   S: Observe.

   ➤ *We can observe with our eyes, but we can also observe by using our other **senses**. We can hear, touch, smell, and sometimes taste. Today we are going to observe something carefully by looking at it, by touching it, and by smelling it, so we are going to _____ .*

   S: Observe.

2. **Introduce** *wood*

   Hold up one of the wood samples. Ask,

   ➤ *Do you know what this is?*

   When students say "wood," write "wood" on the word wall and say,

   *Yes, this little block is made of **wood**. Wood is a kind of **material**.*

   Write "material" on the word wall.

Materials for Step 3
- *Basins of wood samples*

3. **Assemble sets of wood samples**

   Show students the wood samples you have in the clear basins. Tell students that each basin contains samples of a different kind of wood. Explain that each student will get one piece of *each kind* of wood to observe.

   Tell students to walk by the basins and pick up one sample of wood from each basin. When they have all five pieces of wood, students will take the samples to their table and begin to investigate them.

   Call students a few at a time to pick up wood samples and take them to a table. Direct the flow of traffic so that students all move past the basins in the same direction.

4. **Explore wood samples**

   Let students explore the different kinds of wood. It's OK if they want to build little towers, trains, walls, and so on with the samples.

   Visit small groups of students while they work with the wood samples. Focus students' observations by asking questions.

   ➤ *How are these two the **same**? How are they **different**?*

   ➤ *Is any piece of wood heavier than the others?*

   ➤ *How does the wood smell?*

   ➤ *Is the wood **rough** or **smooth**?* (Write "**texture**" on the word wall.)

   ➤ *What color is the wood? Is it all the same color or different colors?*

   ➤ *Is this sample (plywood) made from one piece of wood? How can you tell?*

   ➤ *How do you think this piece of wood (particleboard) was made?*

   ➤ *What else did you find out about wood?* (Point out the lines or patterns on the wood and introduce them as the **grain** of the wood.

5. **Name the wood samples and write the words**

   Call students to the rug *with their wood samples*. Hold up a sample of cedar. Ask students to describe the sample you are holding.

   When students have shared their observations, identify it as **cedar** and write "cedar" on the word wall.

   Ask students to hold up their pieces of cedar and say "cedar."

   Follow this same procedure to identify and name the other four wood samples. Write the names on the word wall.

E L   N O T E

Provide sentence frames if necessary, such as:
It looks _____ .
It feels _____ .
It smells _____ .

Materials for Step 6
- *Tree round*

Materials for Step 7
- *Cedar Tree* poster
- *Pine Tree* poster
- *Linden Tree* poster

Say it
Write it
Hear it
See it
New Word

6. **Discuss the source of wood**
   Ask students,

   ➤ *Where do you think the material wood comes from?*

   Listen to student's prior knowledge. Show students the tree round, and confirm that wood comes from **trees**.

   This is a slice of the trunk of a tree. Can you see something that looks like wood? The tree trunk is all wood except for the thin layer of bark on the outside.

7. **Identify natural wood sources**
   Hold up the piece of cedar and the *Cedar Tree* poster. Tell students,

   This is a piece of cedar. It comes from the cedar tree. This is what a cedar tree looks like.

   Hold up the piece of **pine** wood and the *Pine Tree* poster, followed by the piece of **basswood** and the *Linden Tree* poster. Use the same process to identify the tree that is the source for the wood.

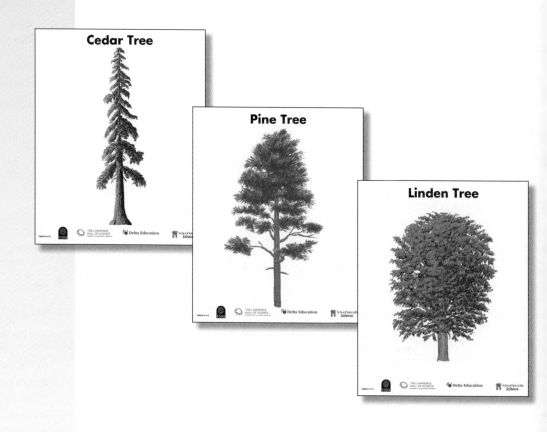

8. Identify human-made wood sources

Hold up a sample of plywood and ask students to hold up their samples. Use several student words or phrases to describe the sample (sandwich wood). Tell students,

➤ *This is called* **plywood**. *Do you think plywood comes from a plywood tree?*

Confirm that there are no plywood trees. Explain that plywood is made by people in factories. Show students the poster of how plywood is made. Discuss the process, step by step.

Repeat the process with **particleboard**, explaining that it is made from wood scraps left over from sawmills. Point out that there are no particleboard trees either. Show the poster of how particleboard is made. Discuss the steps as you review the pictures.

Materials for Step 8
- *Plywood Production* poster
- *Particleboard Production* poster

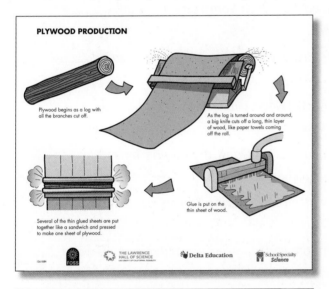

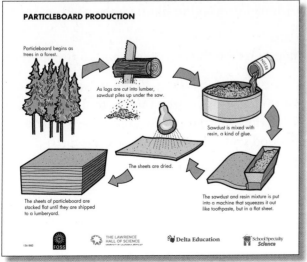

basswood
cedar
different
grain
material
observe
particleboard
pine
plywood
rough
same
senses
smooth
texture
tree
wood

9. **Review vocabulary**

Review key vocabulary, and add the words to the word wall. One way to do this is to use cloze review. Say a sentence, leaving the last word off, and ask students to answer chorally. Here's an example of cloze review for this part.

➤ *When you look at something closely, you _____ .*

S: Observe.

➤ *When you see, touch, hear, or smell something, you are using your _____ .*

S: Senses.

➤ *Wood that comes from a cedar tree is called _____ .*

S: Cedar.

➤ *Wood that comes from a pine tree is called _____ .*

S: Pine.

➤ *Wood that comes from a linden tree is called _____ .*

S: Basswood.

➤ *Layers of wood, glued together, make _____ .*

S: Plywood.

➤ *Wood waste, sawdust, and glue are made into _____ .*

S: Particleboard.

➤ *Light and dark lines in wood are called the _____ .*

S: Grain.

➤ Smooth *and* rough *are words that describe wood _____ .*

S: Texture.

Provide samples of wood to illustrate the above words.

10. **Clean up**

Have students return the pieces of wood to the appropriate basins.

11. **Focus question: Where does wood come from?**

Now that students have learned words to describe wood, experienced different kinds of wood, and learned the names of the trees the wood samples come from, ask the focus question.

➤ *Where does wood come from?*

Write the focus question on the chart, and read it aloud. Tell students you have a strip of paper with the question written on it. Describe and model how to glue the strip into the notebook. Have students use pictures and/or words to answer the question.

12. **Model responses to the focus question**

Depending on students' experience with notebooks, let them work on their own or model a notebook entry, using the focus chart. A focus chart is a sheet of chart paper on which you write the focus question and then ask the class to provide ideas for a response. Model what an answer might look like, and students can then record this response in their notebooks. You might also use the focus chart to provide sentence frames from which students can select and put in their notebooks.

WRAP-UP/WARM-UP

13. **Share notebook entries**

Conclude Part 1 or start Part 2 by having students share notebook entries. Ask students to open their science notebooks to the first entry. Read the focus question together.

➤ *Where does wood come from?*

Ask students to pair up with a partner to

- share their answers to the focus question;
- explain their drawings.

FOCUS CHART

Where does wood come from?

Wood comes from trees.

Pine comes from a
_____ .

Plywood is made by _____ .

E L   N O T E

See the Science-Centered Language Development chapter for strategies on sharing notebook entries.

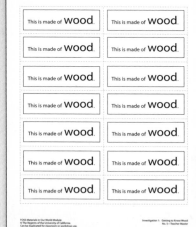

No. 5—Teacher Master

MATERIALS *for*
Part 2: *Wood Hunt*

For each student

2   Wood labels
1   *FOSS Science Resources: Materials in Our World*
  • "The Story of a Chair"

For the class

1   Set of five wood samples (See Step 5 of Getting Ready.)
12   Basswood samples
12   Particleboard samples
12   Pine samples
12   Plywood samples
12   Cedar samples
5   Basins
•   Transparent tape
•   Masking tape
❏   1   Teacher master 5, *Wood Labels*
1   Big book, *FOSS Science Resources: Materials in Our World*

For assessment

•   *Assessment Checklist*

❏   Use the duplication master to make copies.

GETTING READY *for*

Part 2: *Wood Hunt*

1. **Schedule the investigation**

   Part 2 is a whole-class investigation. Plan 20 minutes for the indoor hunt, 15 minutes for the outdoor hunt, and 15 minutes for the reading.

2. **Preview Part 2**

   Students go on a wood hunt and label objects in the classroom that are made of wood. They also go outside to find wood. The focus question is **What is made of wood?**

3. **Prepare for the wood hunt**

   Get six of each kind of wood sample (a total of 30 pieces). While students are at recess or before they come to school, place the samples around the room in obvious locations. There should be one wood piece for each student.

   Put another 30 wood samples (six of each kind of wood) in a basin for distribution to students during the activity. Make sure that you have the same number and kind of samples in this group as those placed around the room—students will be hunting for matching pieces.

   Have one or two extra sets of wood samples ready for students who are unable to find a match in the room. Sometimes, mismatches or "hidden" samples leave students without a sample to find.

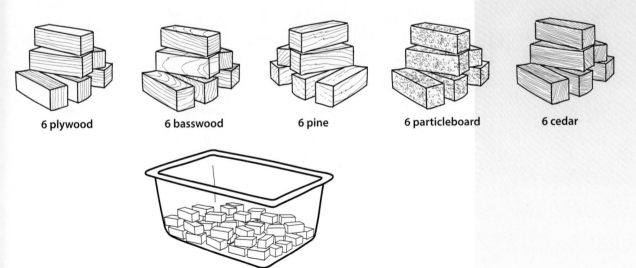

6 plywood   6 basswood   6 pine   6 particleboard   6 cedar

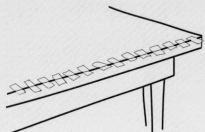

4. **Photocopy the wood labels**

   Make enough copies of teacher master 5, *Wood Labels*, so that each student gets two "This is made of wood." labels. One will be used to label an object in the room and the second to label an object outdoors.

5. **Plan for the introduction**

   Gather a set of wood samples (one of each kind of wood) to use when you introduce the activity.

6. **Plan for tape distribution**

   Students will be using small pieces of transparent tape to attach labels to objects in the classroom made of wood. Getting tape can cause a bottleneck. Put a number of little pieces of tape along the edge of a table or counter for students to get quickly during the activity. Follow a similar procedure for the masking tape used to label wood objects outdoors.

7. **Select your outdoor site**

   In Step 10, you will take students on an outdoor walk to find and label things made from wood. Plan a route that would be appropriate for the activity. Consider the safety of the route.

8. **Check the site**

   Tour the outdoor site on the morning of an outdoor activity. Do a quick check for unsightly and distracting items.

9. **Plan to read** *Science Resources*: **"The Story of a Chair"**

   Plan to read "The Story of a Chair" during a reading period after completing the active investigations for this part.

10. **Plan assessment**

    Here are the specific objectives to observe in this part.

    - *Many objects are made from wood.*

    - *Students compare properties of wood (color, texture, grain, smell).*

    Focus on a few students each session. Record the date and a + or − on the *Assessment Checklist*.

GUIDING *the Investigation*
Part 2: *Wood Hunt*

FOCUS QUESTION
What is made of wood?

1. **Review the wood samples**
   Use a set of the five wood samples. Call students to the rug and review the previous lesson.

   ➤ *What kinds of wood did we investigate?*

   ➤ *How were the wood samples different?*

   ➤ *How were the wood samples the same?*

   ➤ *Where does wood come from?*

2. **Explain the wood hunt**
   Tell students they are going on a wood hunt. Explain that there are wood samples around the room and that each student will get a piece of wood. Tell students their job is to look around the room, find *one* wood sample that is the same as theirs, and bring both pieces back to the rug.

3. **Distribute the wood samples**
   Remind students they should find only one matching piece. Give each student a wood sample. Have students look around the room for a sample that matches. As students return to the rug with their samples, let them share their matches with a partner. If a few students are left at the end of the hunt without finding a match, show them the extra samples you have from which they can choose.

4. **Review the wood names**
   When all students have returned to the rug with their samples, call out the names of the different kinds of wood, and ask students to stand if they found a match for that kind of wood. Ask them which kinds of wood are made by gluing large or small pieces of wood together, and which are cut from trees and don't have to be processed by people.

   basswood
   cedar
   particleboard
   pine
   plywood
   properties

5. **Collect the wood samples**
   Call out the name of one kind of wood, and describe a few of its properties. Tell students,

   We can describe objects by their **properties**. *Things we* **observe** *about objects by looking at them or feeling them are properties of the objects.*

   Ask students with that kind of wood to return their samples to the proper basin.

Say it
Write it New Word See it
Hear it

Materials for Step 7
- *Wood labels*
- *Transparent tape*

6. Focus question: What is made of wood?

Introduce the focus question, and write it on the chart. Read it aloud with students.

➤ *What is made of wood?*

Ask students to look around the room to find things that are made of wood.

7. Label wood objects in the classroom

Tell students to find one object in the room that is made of wood and label the object, using the following procedure. Model the process as you give the directions.

a. Pick up a "This is made of wood." label and a piece of tape.

b. Label something made of wood in the room.

c. Return to the rug.

8. Observe wood in class

When all students have returned to the rug, have them look around the room and take note of all the things that are made of wood. Discuss how students can tell when something is made of wood and why they think wood is a good material for making the things they found. [It's sturdy, strong, flat, can be cut, can be painted.]

Materials for Step 9
- *Wood labels*
- *Masking tape*

9. Search outdoors for wood

Ask students if they might be able to find objects made of wood outside. Give each student another "This is made of wood." label and a small piece of masking tape.

NOTE: These labels will remain outdoors so students can observe the labels in Investigation 3 when they investigate paper.

10. Go outdoors

Walk your selected route outdoors to look for wood objects. Students should use what they know about wood and their observation skills to figure out if something is made of wood.

When every student has had an opportunity to label a wood object, gather students in a circle. Explain what they will do next. When you give your signal to go, each student should walk to one of the objects made of wood and stand by it. Give the signal and have students move out. Make sure that all students are touching a wood object. Repeat this procedure once or twice.

Some schoolyards contain synthetic lumber, a composite material made of recycled plastic and wood. Have students observe this material to decide if it is wood.

TEACHING NOTE

Make a list of what students label outside for the discussion indoors.

Students learned in Part 1 that wood is made from trees. See if they label trees.

11. Return to class

In an orderly manner, return to the classroom for a discussion. Gather students at the rug and ask them to share what they found outside that was made of wood.

On the word wall, write the names of the objects that students offer, and add any additional objects you had on your list. Ask students to compare a wood object found outside to one found inside. Ask students how the objects are the same and how they are different.

12. Answer the focus question

Restate the focus question, and have the class read it aloud together.

➤ *What is made of wood?*

Tell students you have a strip of paper with the question written on it. Describe and model how to glue the strip into the notebook.

Ask students to select one of the objects in the room or outside that is made of wood and to draw it in their notebooks. Tell students to caption their drawings by writing "<Object> is made of wood." Write this sentence frame on the focus chart, and model how students would use it to caption their drawings.

FOCUS CHART

What is made of wood?

_____ is made of wood.

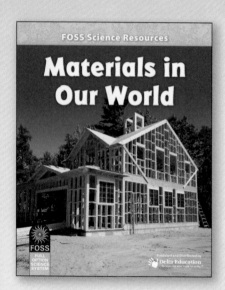

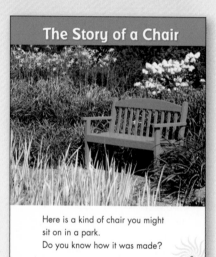

Here is a kind of chair you might
sit on in a park.
Do you know how it was made?

3

READING *in Science Resources*

13. Read "The Story of a Chair"

Give students a few minutes to look at and discuss the cover of the
book. Ask students to find the wood in the photo.

Ask students what objects they found in the classroom that were
made of wood. Continue the discussion by asking where the wood
came from to make the items. Introduce "The Story of a Chair."
Tell students that this article will tell them how a chair is made
from wood.

Read aloud "The Story of a Chair." Pause to discuss key points
in the article, to review the pictures, and to make predictions.
For reading strategies to support English learners, see the
Science-Centered Language Development chapter.

14. Discuss the reading

Discuss the article, using these questions as a guide.

➤ *Where did the wood come from to make the chair?*

➤ *How does the wood from a tree become a chair?*

➤ *Why is wood a good material to use to make a chair?*

15. Extend the reading

Use this activity to deepen students' understanding after students
have reread the article.

In small groups, have students use the format of "The Story of a
Chair" to dictate their own story of a wooden object. For example,
one group could create the story of a table, another group the story
of a baseball bat, and a third group the story of a pencil. Write the
stories on chart paper, and have each student illustrate a portion of
the group's story. Post the completed stories, and, as a class, discuss
the similarities among them.

WRAP-UP/WARM-UP

16. Share notebook entries

Conclude Part 2 or start Part 3 by having students share notebook entries. Ask students to open their science notebooks to the most recent entry. Read the focus question to the class.

➤ *What is made of wood?*

Ask students to pair up with a partner to

- share their answers to the focus question;

- explain their drawings.

No. 6—Teacher Master

MATERIALS *for*
Part 3: *Wood and Water*

For each pair of students at the center

- 2   Droppers
- 1   Plastic cup
- 2   Sets of wood samples (See Step 6 of Getting Ready.)
- 1   Basin

For the class

- 5   Containers, 1/2 L
- •   Sponges
- •   Water ★
- 1   Pitcher or empty 2 L soft-drink bottle ★
- •   Paper towels ★
- •   Newspaper ★
- 1   Turkey baster (optional) ★
- 1   Bag for carrying materials outdoors ★
- ❑   1   Teacher master 6, *Center Instructions—Wood and Water*

For assessment

- •   *Assessment Checklist*

★ Supplied by the teacher.   ❑ Use the duplication master to make copies.

GETTING READY *for*
Part 3: *Wood and Water*

1. **Schedule the investigation**

   This part requires 15–20 minutes at the center for each group of six to ten students and 15 minutes for the class outdoors. It can also be done as a whole-class activity. Do this entire part outdoors if you have a suitable place.

   In addition, plan 10 minutes to introduce, 15 minutes for recording in notebooks, and 5 minutes to wrap up the session with the entire class at the rug.

2. **Preview Part 3**

   Students observe how wood and water interact, first by putting drops of water on the wood, then by putting the wood in basins of water. The focus question is **What happens when wood gets wet?**

3. **Put water in cups and basins**

   For each pair of students, fill a cup about one-third full of water and a clear basin about half full of water. The water should stay clean enough to use with several groups. You may have to add a little water to the cups after each group.

   As an alternative to using a basin for each pair of students, you can use a smaller 1/2-liter (L) container. Using 1/2 L containers will allow you to do this part as a whole-class activity.

4. **Plan for the introduction**

   Have one set of wood samples ready to introduce the activity. If students haven't used droppers before, be prepared to demonstrate with a cup of water and a turkey baster. (See Step 4 in Guiding the Investigation.)

5. **Plan for cleanup**

   Keep in mind the possibility of a water spill. Know in advance where to get a mop, and keep sponges and paper towels handy.

   Plan a space in the classroom for the wood samples to dry overnight on newspaper. The wood samples should be completely dry before storing them in the kit.

6. **Set up the center**

   Put a dropper at each student's place and a cup of water for each pair of students to share. Keep basins of water nearby so they can be easily moved to the center when needed.

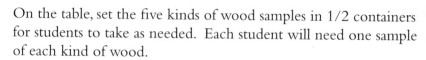

On the table, set the five kinds of wood samples in 1/2 containers for students to take as needed. Each student will need one sample of each kind of wood.

Have the class word wall near the center so you can add vocabulary to it while students work. Make sure it's handy every time you use the center.

7. Select your outdoor site

In Step 15, you will take students on an outdoor walk to find out what happens when outdoor wood gets wet. Plan a route that would be appropriate for the activity. Consider the safety of the route.

8. Check the site

Tour the outdoor site on the morning of an outdoor activity. Do a quick check for unsightly and distracting items.

9. Plan assessment

Here are the specific objectives to observe in this part.

- *Wood has observable properties (color, pattern, grain). When it gets wet, wood looks darker.*

- *Different kinds of wood absorb water in different ways. Water sometimes forms beads on the wood; sometimes water soaks into the wood and spreads out.*

- *Students conduct a simple investigation.*

Focus on a few students each session. Record the date and a + or − on the *Assessment Checklist*.

GUIDING *the Investigation*
Part 3: *Wood and Water*

1. **Introduce** *communicate*

   Call students to the rug. Introduce the word *communicate* with a mini-lesson. One way to do this is to have students respond chorally. Below is an example of this mini-lesson.

   ➤ *Today we are going to learn another very important science word. Say "**communicate**."* (Write the word on the word wall.)

   S: Communicate.

   ➤ *Let's clap the parts (syllables) of the word.*

   S: Com (clap) mun (clap) i (clap) cate (clap).

   ➤ *Again.*

   S: Com (clap) mun (clap) i (clap) cate (clap).

   ➤ *How many parts (syllables) are in the word* communicate?

   S: Four.

   ➤ *Communicate means to tell about something by talking or writing. What does communicate mean?*

   S: To tell about something by talking or writing.

   ➤ *When I tell about something by talking or writing, I _____ .*

   S: Communicate.

2. **Review the wood samples**

   Review the wood samples students have been exploring. Tell them that today they will investigate what happens when they drop water onto wood. Tell students that they will use cedar and particleboard samples. Ask students to guess what will happen.

3. **Focus question: What happens when wood gets wet?**

   Write the focus question on the chart, and read it aloud.

   ➤ *What happens when wood gets wet?*

4. **Demonstrate the dropper (optional)**

   If students have not worked with droppers before, show them how to use one. A turkey baster makes a good demonstration tool.

   a. *Squeeze the rubber bulb between your thumb and first finger.*

   b. *Lower the tip of the dropper into the cup of water.*

   c. *Release the pressure on the bulb, keeping the tip under the water.*

FOCUS QUESTION
What happens when wood gets wet?

TEACHING NOTE

These two samples, cedar and particleboard, were chosen because water soaks into them at very different rates.

Materials for Step 4
- *Cup of water*
- *Dropper*
- *Turkey baster (optional)*

d. *Lift the dropper out of the water, being careful not to squeeze the bulb.*

e. *Move the dropper into position and gently squeeze the bulb to release the water, drop by drop.*

5. **Drop water on cedar and particleboard**

Send six to ten students to the center. Pass the containers of cedar and particleboard around the group. Have each student take one sample of each kind of wood. Suggest that students put one drop of water on the surface of each sample, then carefully observe what happens. Next, have them put a drop of water on the end of each wood sample and observe.

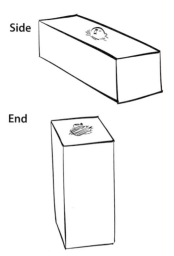

Materials for Step 5
- *Cups of water*
- *Droppers*
- *Containers of wood samples*
- *Basins of water*
- *Sponges or paper towels*
- *Pitcher of water*

Say it → New Word → See it → Hear it → Write it

6. **Monitor progress and discussion**

Encourage students to discuss what happens when they drop water on the wood samples. Guide their observations by asking questions.

➤ *Does water sit on the pieces of wood, or does it **soak** in?*

➤ *How does water soak in? (Does it get **absorbed** right away, or does it take some time?)*

➤ *Does the color of the wood change when it gets wet?*

➤ *Does it matter if you put the drop on the side or on the end of the sample?*

➤ *What is the shape of the water when it sits on a piece of wood? Is the water flat, and does it **spread** out or does it form a **bead**?*

Allow plenty of time for exploration, but call the activity to a close if students start to splash water too much.

7. Add vocabulary to the word wall

As students offer their observations, add any new or important vocabulary to the word wall. Let students be the guides—acknowledge the words they use, and offer new vocabulary as needed, such as *absorb*, *bead*, and *spread*.

8. Investigate the other wood samples

Let students continue investigating. Have them use the other wood samples when they are ready.

9. Use basins of water

Ask students what they think will happen if they put the wood samples in a basin of water. Remove the cups of water and droppers and put the basins of water on the table so that each pair shares one. Have students put their samples of wood into the water one at a time. (All the samples should float.)

TEACHING NOTE

Students often discover that when the samples get completely soaked, they stick together. Challenge students to see how many samples will stick together when students pick them up.

10. Introduce *sink* and *float*

To introduce **sink** and ***float***, gather two or three small objects that will float and two or three objects that will sink in the water. When you introduce the lesson, put each object in the water, one at a time, and discuss whether it sinks or floats.

11. Clean up

After students have had ample time to explore with the wood and water, remove the basins from the table, but keep them handy for the next group. Spread the wood samples on newspaper to dry. Be sure to let the wood dry before storing.

Say it
Write it · New Word · See it
Hear it

absorb
bead
communicate
float
sink
soak
spread

12. Review vocabulary

Review key vocabulary added to the word wall. Here's a suggested cloze review. Students answer chorally.

➤ *When I tell about something by talking or writing, I _____ .*
S: Communicate.

➤ *When wood stays on top of the water, we say it _____ .*
S: Floats.

➤ *When the wood goes to the bottom of the water, we say it _____ .*
S: Sinks.

➤ *When the wood absorbs the water, we say the water _____ in.*
S: Soaks.

Review other words that came up during the activity.

- Absorb

- Bead

- Spread

B R E A K P O I N T

13. Discuss water on wood

Gather students at the rug. Review what happened when the wood samples got wet by using drops of water. Ask students to share their ideas.

Explain that students will have a chance to go outside and find out what happens when outdoor wood gets wet. Explain that they will work with a partner and use a dropper to drop water on outdoor wood and observe what happens.

14. Distribute cups and droppers

Tell students that one partner will carry the dropper, while the other partner carries the empty cup. Distribute the materials.

15. Go outdoors

Pair students up in a line and head outdoors in the usual orderly manner. Gather the group in your outdoor meeting spot and review the boundaries for this activity. Make sure students remember what their task is and how to work with a partner. Have students test three to five wood objects. Quickly review the rules for outdoor work, give each pair of students a bit of water, and let the water dropping begin.

Materials for Step 14
- *Cups*
- *Droppers*
- *Bottle of water*

As students work, encourage them to seek out different kinds of wood—rotten wood, painted wood, sealed wood, and untreated wood.

As they work, ask students,

➤ *What happens to the water on this wood?*

➤ *Does the water soak into the wood or bead up on the wood?*

➤ *Is the wood covered with anything?*

➤ *Does the wood look like it is new or old?*

➤ *Why do you think this object does or doesn't absorb water?*

16. Return to class

In an orderly manner, return to the classroom and gather students at the rug. Ask them to share a few things they found out about wood outdoors. Guide the conversation so that students consider that outdoor wood is often protected by paint or sealed with something to protect it from rain and moisture.

17. Answer the focus question

Restate the focus question.

➤ *What happens when wood gets wet?*

Tell students you have a strip of paper with the question written on it. Review how to glue the strip into the notebook.

Ask students to answer the question in pictures and/or words. You can write sentence frames on the focus chart and model how students would use the frames to caption their drawings.

FOCUS CHART

What happens when wood gets wet?

The wood looks
——— .

The water ——— .

WRAP-UP/WARM-UP

18. Share notebook entries

Conclude Part 3 or start Part 4 by having students share notebook entries. Ask students to open their science notebooks to the most recent entry. Read the focus question together as a class.

➤ *What happens when wood gets wet?*

Ask students to pair up with a partner to

• share their answers to the focus question;

• explain their drawings.

CENTER INSTRUCTIONS—SINK THE PINE AND PLYWOOD

Materials

Basins half filled with water
Jumbo paper clips in cups
Rubber bands in a cup
Sponges
Pine samples
Plywood samples
Paper towels
Newspaper

Set Up the Center

Set a clear basin half filled with water on the table for each pair of students. Put cups of jumbo paper clips near each basin and a cup of rubber bands in the center of the table within easy reach of all students. Have pine and plywood samples ready.

Guide the Investigation

1. **Discuss sinking.** Show students the paper clips and rubber bands. Ask students to see if they can use these materials to sink the two kinds of wood.
2. **Begin the exploration.** Distribute one sample of pine and plywood to each student. Let students explore freely. (If students are stumped, show them how to put a rubber band around a piece of wood and slide a paper clip under the rubber band.)
3. **Quantify the results.** When students can attach paper clips to a sample, challenge them to find out how many paper clips are required to sink each piece of wood.

   NOTE: Young students may not use a systematic method to determine how many paper clips it takes to sink the wood. They love to attach as many paper clips as they can. Let them use their own methods, regardless of accuracy.
4. **Ask questions to guide discussion.**
   - *Were you able to sink the wood by attaching paper clips?*
   - *How many paper clips did you use to sink the wood?*
   - *Let's compare the two samples. Does it take the same number of paper clips to sink both kinds of wood?*
   - *Does it make a difference where you put the paper clips on the wood? All on one side? Evenly distributed around all sides?*
5. **Add to the word wall.** Add new or important words that students use to the word wall.
6. **Clean up.** Have students remove the paper clips and rubber bands and return them to their containers for the next group. Lay the wet wood samples on newspaper to dry. After all the groups are finished, spread the paper clips on towels to dry.

Vocabulary

Try to include these words in discussions with students:
above, below, compare, float, sink, test

FOSS Materials in Our World Module
© The Regents of the University of California
Can be duplicated for classroom or workshop use.

Investigation 1: Getting to Know Wood
No. 7—Teacher Master

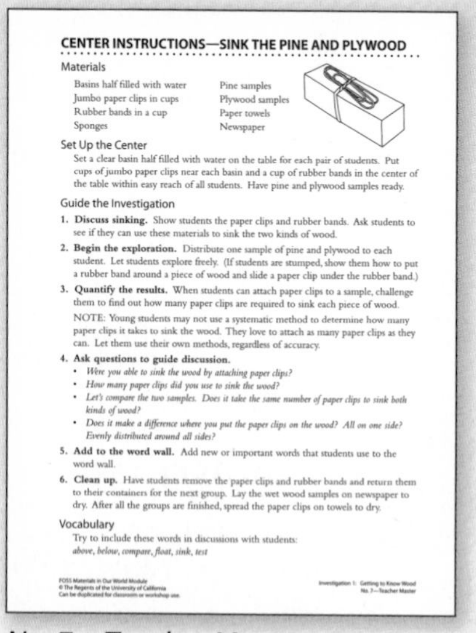

No. 7—Teacher Master

MATERIALS *for*

Part 4: *Sink the Pine and Plywood*

For each pair of students at the center

- 1 Basin
- 2 Containers, 1/2 L (Optional; see Step 3 of Getting Ready.)
- 2 Pine samples
- 2 Plywood samples
- 1 Plastic cup
- 40 Paper clips, jumbo
- 4–6 Rubber bands, #8 (50–60 in one cup for all students to share)

For the class

- 1 Set of five wood samples
- 1 Plastic cup
- • Sponges
- • Water ★
- • Paper towels ★
- • Newspaper ★
- ❑ 1 Teacher master 7, *Center Instructions—Sink the Pine and Plywood*

For assessment

- • *Assessment Checklist*

★ Supplied by the teacher.

❑ Use the duplication master to make copies.

GETTING READY *for*

Part 4: *Sink the Pine and Plywood*

1. **Schedule the investigation**

   This part requires 20–30 minutes at a center for each group of six to ten students. It can also be done as a whole-class activity. Do this entire part outdoors if you have a suitable place.

   Plan 10 minutes to introduce, 15 minutes for recording in notebooks, and 5 minutes to wrap up.

   Read ahead to Part 5. If you feel that your students are ready for it, you can move on to the Sinking Investigation with the whole class or as a challenge for some pairs of students.

2. **Preview Part 4**

   Students find ways to sink two of the floating wood samples by attaching paper clips to the wood with rubber bands. Students discover how easy it is to sink the plywood compared to the pine sample. They use relative-position words to describe the location of the wood. The focus question is **How can you sink wood?**

3. **Fill basins with water**

   Half fill a basin with water for each pair of students. The water should stay clean enough that you will not have to refill the basins after each group.

   As an alternative to using a basin for each pair of students, you can use a smaller 1/2 L container. Using 1/2 L containers will allow you to do this part as a whole-class activity.

4. **Prepare paper clips and rubber bands**

   Put about 40 jumbo paper clips into a plastic cup for each pair of students. Put 50–60 rubber bands in a cup for all students at the center to share.

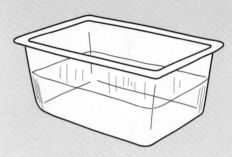

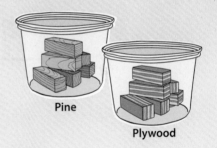
Pine

Plywood

▶ **NOTE**

Be aware of any allergies your students have, including latex allergies. (Rubber bands are made of latex.)

5. **Prepare for the introduction**

   You'll need one of the basins of water and a set of wood samples to introduce the lesson.

   If you feel that students need to review the concepts of sinking and floating, you may want to begin with a lesson on what the words *sink* and *float* mean.

   To review sink and float, gather two or three small objects that will float in water and two or three objects that will sink. When you introduce the lesson, put the objects in water one at a time, and discuss whether they sink or float.

6. **Set up the center**

   Set the basins filled with water on the table. Each pair of students will use one basin (or one 1/2 L container as an alternative). Put cups of paper clips near each basin and the cup of rubber bands in the center of the table so they will be within easy reach of all students. Have the pine and plywood samples ready for distribution. Be prepared to clean up spills with sponges and paper towels.

7. **Plan assessment**

   Here are the specific objectives to observe in this part.

   * *Some kinds of wood sink more easily than others.*

   * *Students compare the properties of wood.*

   * *Students conduct a simple investigation.*

   Focus on a few students each session. Record the date and a + or – on the Assessment Checklist.

GUIDING *the Investigation*
Part 4: *Sink the Pine and Plywood*

1. Introduce *compare*
Call students to the rug. Introduce the word *compare* with a mini-lesson. One way to do this is to have students respond chorally. Below is an example of this mini-lesson.

➤ *Today we are going to learn a very important science word. Say "compare."* [Write the word on the word wall.]

S: Compare.

➤ *Let's clap the parts (syllables) of the word.*

S: Com (clap) pare (clap).

➤ *Again.*

S: Com (clap) pare (clap).

➤ *How many parts (syllables) are in the word* compare?

S: Two.

➤ *Compare means to look at two different things and observe how they are the same and different. What does compare mean?*

S: To look at two different things and observe how they are the same and different.

➤ *When I look at two different things and tell how they are the same and different, I _____.*

S: Compare.

2. Review wood and water
Ask students what happened when they put the wood samples in the basins of water. [They all floated.] Review the meaning of the words *sink* and *float* if necessary.

3. Focus question: How can you sink wood?
Write the focus question on the chart, and have students read it together.

➤ *How can you sink wood?*

4. Describe sinking wood
Tell students that their new challenge will be to sink the pieces of wood. Explain that each student will get a piece of pine and a piece of plywood to work with. Discuss ways that they might be able to sink the wood pieces. Describe this process as a **test**. Tell students that they will be testing to find out how to sink the pine and the plywood.

Materials for Step 5
- *Cup of rubber bands*
- *Cups of paper clips*
- *Containers of pine samples*
- *Containers of plywood samples*
- *Basins of water*
- *Sponges and paper towels*

▶ **SAFETY NOTE**

Be aware of any allergies your students have, including latex allergies. (Rubber bands are made of latex.)

TEACHING NOTE

Many early-childhood students will not use a systematic method to determine how many paper clips it takes to sink the wood. They love to stick as many paper clips on as they can. That's OK for this part of the activity.

---

5. **Send students to the center**

Send six to ten students to the center. Show them the paper clips and rubber bands. Ask them to see if they can use these materials to sink the two kinds of wood. Remind students that this is called a test.

6. **Explore sinking**

Distribute samples of pine and plywood to each student. Let students start a free exploration of sinking.

If students are stumped, show them how to place a rubber band around a piece of wood and slide a paper clip under the rubber band.

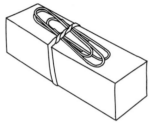

7. **Quantify results**

When students have mastered a technique for attaching paper clips to wood, challenge them to find out how many paper clips it takes to sink each piece.

Set a standard for "sink." The piece is not sunk until it lies flat on the bottom of the basin. Sometimes one end sinks, but the other end floats up. The paper clips must be in the middle of the wood to allow the wood to float evenly. This is a good time to reinforce descriptions of relative position of the wood with respect to the bottom of the container or the surface of the water (e.g., the wood is **below** the surface of the water; the wood is **above** the bottom of the container).

8. **Ask questions to guide discussion**

Ask questions to guide students' observations and discussion.

➤ *Were you able to sink the wood by attaching paper clips?*

➤ *How many paper clips did you use to sink the wood?*

➤ *Let's compare the two samples. Does it take the same number of paper clips to sink both kinds of wood?*

➤ *Does it make a difference where you put the paper clips on the wood? All on one side? Evenly distributed around all sides?*

9. **Add vocabulary to word wall**

As students offer their observations, add any new or important vocabulary to the word wall. Let students be the guides—acknowledge the words they use, and offer new vocabulary as needed.

10. **Clean up**

To clean up the station for the next group, have students take all the paper clips and rubber bands off the wood samples and put them back in their containers. Lay out the wet wood samples on a sheet of newspaper or in a basin to dry.

Dry the paper clips by spreading them out on paper towels at the end of the part. Make sure the clips are completely dry before putting them back in the box or plastic bag.

Recycle or reuse the water by watering some plants.

11. **Review vocabulary**

Review key vocabulary added to the word wall. Here's a suggested cloze review. Students answer chorally.

above
below
compare
float
sink
test

➤ *When I look at two different things and tell how they are the same or different, I _____ .*

S: Compare.

➤ *We attached paper clips to wood in different ways and amounts in order to sink the wood. This is called a _____ .*

S: Test.

Discuss these additional words, and write them on the word wall.

- Above
- Below

12. **Answer the focus question**

Restate the focus question.

➤ *How can you sink wood?*

Tell students you have a strip of paper with the question written on it. Review how to glue the strip into the notebook.

Ask students to answer the focus question in pictures and/or words and show how they were able to sink the wood.

FOCUS CHART

How can you sink wood?

You can sink wood by _____ .

EL NOTE

See the Science-Centered Language Development chapter for strategies on sharing notebook entries.

WRAP-UP/WARM-UP

13. Share notebook entries

Conclude Part 4 or start Part 5 by having students share notebook entries. Ask students to open their science notebooks to the most recent entry. Read the focus question together as a class.

➤ *How can you sink wood?*

Ask students to pair up with a partner to

- share their answers to the focus question;
- explain their drawings.

MATERIALS *for*

Part 5: *Sinking Investigation (Optional)*

For each pair of students at the center

2	Particleboard samples
2	Cedar samples
1	Basin
1	Plastic cup
30	Paper clips, jumbo
4–6	Rubber bands, #8 (50–60 in one cup for all students to share)

For the class

- 1 Pine sample
- 1 Plastic cup
- • Sponges
- • Water ★
- • Paper towels ★
- • Newspaper ★
- ❏ 1 Teacher master 8, *Center Instructions—Sinking Investigation*

For assessment

- • *Assessment Checklist*

★ Supplied by the teacher.   ❏ Use the duplication master to make copies.

No. 8—Teacher Master

GETTING READY *for*
Part 5: *Sinking Investigation (Optional)*

1. **Schedule the investigation**
   This part requires 20 minutes at the center for each group of six to ten students. It can also be done as a whole-class investigation if students work in pairs. Plan 10 minutes to introduce and 15 minutes for students to record in their notebooks.

2. **Preview Part 5**
   Students refine their techniques of using paper clips and rubber bands to sink wood samples, and they test two kinds of wood. They make a bar graph of their results, lining up and comparing the number of clips used. The focus question is **How many paper clips does it take to sink wood?**

3. **Make a decision**
   This part works best when students have more experience working with materials. If you are doing this module early in the fall, you may want to return to this part later in the year when students can deal with being more systematic.

4. **Prepare paper clips and rubber bands**
   Put about 30 paper clips into a plastic cup for each pair of students. Put about 50 rubber bands in a cup for all students to share.

5. **Plan for the introduction**
   You'll need a basin of water, a sample of pine, ten paper clips, and a rubber band to introduce the investigation.

6. **Set up the center**
   Set a basin half filled with water on the table for each pair of students. As an alternative to using a basin for each pair of students, you can use a smaller 1/2 L container. Using 1/2 L containers will allow you to do this part as a whole-class activity.

   Put cups of paper clips near each basin (or 1/2 L container) and one cup of rubber bands in the center of the table. Have the cedar and particleboard ready for distribution.

7. **Plan assessment**
   Here are the specific objectives to observe in this part.

   - *Some kinds of wood sink more easily than others.*

   - *Students conduct a simple investigation.*

   - *Students organize observations.*

Cedar

Particleboard

GUIDING *the Investigation*
Part 5: *Sinking Investigation (Optional)*

FOCUS QUESTION
How many paper clips does it take to sink wood?

1. **Review Part 4**

   Call students to the rug and ask if they were able to sink the wood samples last time they worked with these materials. Ask how they did it. [Attached paper clips with rubber bands to the pieces of wood.]

2. **Introduce sinking challenge**

   Tell students that today they are going to work with two pieces of wood, and that this time they are going to find out exactly how many paper clips it takes to sink each piece.

3. **Focus question: How many paper clips does it take to sink wood?**

   Write the focus question on the chart, and have students read it together.

   ➤ *How many paper clips does it take to sink wood?*

4. **Demonstrate the test with pine**

   Use a pine sample to demonstrate the procedure that students are to follow at the center.

   a. *Put a rubber band around the sample.*

   b. *Slip one paper clip under the rubber band, and place it in the water to see if the pine continues to float.*

   c. *Slip another paper clip under the rubber band on the pine sample. Test to see if it still floats.*

   d. *Continue this procedure, testing after you add each paper clip, until the pine sample sinks. [It should take six to eight jumbo clips to sink the pine sample.]*

   e. *Set the piece of wood aside with the rubber band and paper clips still in place (so students have data to make a graph with later).*

5. **Send students to the center**

   Send six to ten students to the center. Have them follow the procedure outlined in Step 4 for each sample of wood: cedar and particleboard. Monitor progress to make sure students test after they add each paper clip and that they leave the paper clips on the samples once students have sunk the samples.

Materials for Step 5
- *Cup of rubber bands*
- *Cups of paper clips*
- *Container of cedar samples*
- *Container of particleboard samples*
- *Basins of water*
- *Sponges and paper towels*

Say it
Write it
New Word
See it
Hear it

6. Create bar graphs

Move the water basins to another location where they will be out of the way, but easily retrieved for use with the next group. This will give you more room for the graphs and help students avoid the temptation of playing in the water.

Have students take the paper clips off their cedar samples. Have them place their pieces of cedar on the table and lay the paper clips beside or below it to start a bar **graph**. Follow the same procedure for the other wood sample.

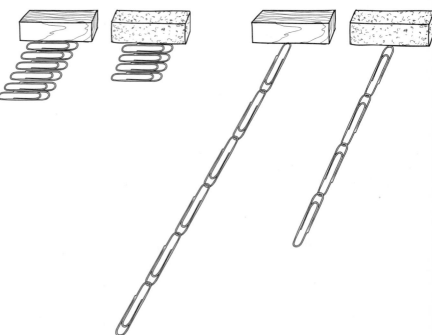

These two illustrations show two ways to graph. Which graph students choose to use will depend on how much room they have.

7. Discuss bar graphs

When students have finished making their graphs, discuss the results. Ask,

➤ *Which piece of wood took **more** paper clips to sink, the cedar or the particleboard?* [Usually the cedar.]

➤ *How could you tell without counting paper clips?* [Find the longer line of clips.]

➤ *Which piece of wood took **fewer** paper clips to sink? (Fewer is another word for **less**.)*

➤ *How many paper clips did it take to sink the cedar? The particleboard?*

fewer
less
more
graph

8. **Add vocabulary to word wall**

   As students offer their observations, add any new or important vocabulary to the word wall. Let students be the guides— acknowledge the words they use, and offer new vocabulary as needed.

9. **Clean up**

   When all groups have finished, dry all equipment thoroughly before returning it to the kit. Spread the paper clips on newspaper, pat them dry with a paper towel, and let them dry overnight to prevent rust. Spread the wood on towels or newspaper to dry overnight.

10. **Answer the focus question**

    Restate the focus question.

    ➤ *How many paper clips does it take to sink wood?*

    Tell students you have a strip of paper with the question written on it. Review how to glue the strip into the notebook.

    Ask students to answer the focus question in pictures and/or words and show how they were able to sink the wood.

> **FOCUS CHART**
>
> *How many paper clips does it take to sink wood?*
>
> It took _____ paper clips to sink particleboard.
>
> It took _____ paper clips to sink cedar.

WRAP-UP

11. **Share notebook entries**

    Conclude Part 5 by having students share notebook entries. Ask students to open their science notebooks to the most recent entry. Read the focus question together as a class.

    ➤ *How many paper clips does it take to sink wood?*

    Ask students to pair up with a partner to

    - share their answers to the focus question;
    - explain their drawings.

TEACHING NOTE

Refer to the teacher resources on FOSSweb for a list of appropriate trade books that relate to this module.

INTERDISCIPLINARY EXTENSIONS

Language Extensions

- **Create a wood chart**

  Create a chart to use throughout the **Materials in Our World Module**. After you discuss the names of the woods, set up the chart. On the chart, tape a sample of each kind of wood, and write the name of the wood under each sample. Then record a few words, phrases, or sentences dictated by students that tell what students know about wood. Use the chart for review, and add to it as the module progresses. If students draw in their science notebooks, the words on the display will also be available for them to copy.

- **Create sorting challenges**

  Write a sorting challenge on a sheet of paper. Here are some possibilities.

  - Sort these objects by color.

  - Sort these objects by size.

  - Sort these objects by shape.

  - Sort these objects by texture.

  Provide a basket of objects for students to sort. One week provide pattern blocks, another week provide buttons, another week provide crayons, and so on.

- **Create a class book of objects made of wood**

  Have each student contribute a page to a class book called *Objects Made of Wood*. Explain that each student's page will consist of a drawing of the object and a sentence caption, such as "A chair is made of wood."

Math Extensions

- **Weigh paper clips**

  Reinforce the concept that students are adding mass (students may say *weight*) to the wood to make it sink. (Some students perceive paper clips as being so light that the paper clips don't weigh anything.) Use a balance to demonstrate that paper clips do have mass.

- ### List wooden items from home
  Have students work with their families to make lists of items at home that are made out of wood. Use chart paper to collapse students' lists into a comprehensive class list or into a table of items grouped by type (tool, furniture, toy, other). Have students count the number of items they found. (See the Home/School Connection on the next page.)

Social Studies Extensions

- ### Visit a lumberyard
  Contact a lumberyard and plan a field trip. Look at the different kinds of wood available and all the different shapes and sizes the wood is cut into. While you are there, ask a yard person to saw a piece of wood and point out the sawdust produced. Ask him or her to demonstrate the jointer, too, if the yard has one. Shavings come from the jointer.

  This is an opportunity to introduce students to careers in construction that involve wood and lumber.

- ### Visit a construction site
  Take a field trip to new housing under construction or another site where wood is being used to build a structure. Talk to the carpenter about how wood is used to make sturdy buildings.

Science Extensions

- ### Observe inside a branch
  Cut a length of a tree branch (bark and all) in half, the long way, so it can be laid open to see the wood inside. Take students on a tree walk to think about all the wood that might be in a tree. Bring the cut branch so students can see what a branch or trunk looks like when it is cut open to reveal the wooden interior.

- ### Start a wood center
  Plan to have a small table designated as a wood center while you are doing the activities in this module. You might start with the tree and wood-production posters and the wood samples that go with each. Borrow the colored posters, *Trees* book, loupe/ magnifying lens, and tree rounds from the **Trees and Weather Module**, if you have one available.

  Bring in interesting pieces of local woods and a variety of things made out of wood. Add any wood pieces that students bring from

> **TEACHING NOTE**
>
> *Review the teacher resources on FOSSweb for community- and career-related extensions.*

> **TEACHING NOTE**
>
> *Review the online activities for students on FOSSweb for module-specific science extensions.*

home to share. Another week, add things made from wood so that students can continue to explore by matching pairs, making patterns, and so on. These might include wood blocks, toothpicks, wood-pattern blocks, wooden marble tracks, and the like. Let this be a free-choice station for students to explore.

- **Play a memory game**
  Use two samples of each kind of wood provided in the kit (ten pieces) and ten paper cups large enough to cover the pieces so they can't be seen. Lay out the wood pieces and cover each with a paper cup. Mix the cups around before beginning. Play a memory game by lifting two cups at a time to find a match (the same kind of wood). When a match is found, the student gets to collect the pieces. If the pieces don't match, cover them back up and move on to the next student's turn.

- **Conduct another sinking-wood investigation**
  Ask students to find out if it makes a difference where the paper clips are placed. Have students answer the question: Does it take more, fewer, or the same number of clips to sink the wood if the clip are placed evenly around the sides of the wood sample or placed all on one side?

TEACHING NOTE

Families can get more information about Home/School Connections on FOSSweb.

Home/School Connection

The sheet for the home/school connection has four labels on it that students can use to label objects at home. It also has four places to draw or list the objects students find. Families are encouraged to give clues: "I'm thinking of something that is made of wood, and it is _____ ."

Make copies of teacher master 9, *Home/School Connection* for Investigation 1. Send it home with students after Part 2.

No. 9—Teacher Master

Investigation 2: Changing Wood

PURPOSE

Content

- Wood can be changed (appearance and behavior) by mechanical action, such as sanding and mixing with water.

- Sawdust is tiny pieces of wood. Sawdust can be recycled into usable wood.

- Basic materials can be transformed into new materials (particleboard and plywood).

- Mixtures are formed when two or more materials are put together.

- Liquid water left in an open container dries up (evaporates).

Scientific Practices

- Observe common objects by using the senses.

- Describe properties of common objects.

- Communicate observations orally and through drawings.

- Explore the technology of making wood products.

INVESTIGATION 2 – *Changing Wood*

Investigation Summary	Time	Focus Question
PART 1 — **Sanding Wood** Students add to their knowledge of the properties of wood and learn how to use those properties to change wood. Students use sandpaper to change the shape of basswood and a stick.	**Introduction** 5 minutes **Outdoors or Center** 30 minutes **Notebook** 20 minutes	How can you change the shape of wood?
PART 2 — **Sawdust and Shavings** Students compare sawdust and shavings. Students find out what happens to sawdust and shavings when they mix the two with water and then separate out the shavings. Students spread out wet sawdust on paper plates and put some in a closed container.	**Introduction** 5 minutes **Center** 20 minutes **Notebook** 20 minutes **Follow-up** 10 minutes	How are sawdust and shavings the same? How are sawdust and shavings different?
PART 3 — **Making Sawdust Wood** Students simulate the making of particleboard by using sawdust and a cornstarch matrix. Students compare their particleboard with the samples from the kit.	**Introduction** 10 minutes **Center** 20 minutes **Notebook** 20 minutes **Follow-up** 10 minutes	How is particleboard made?
PART 4 — **Making Sandwich Wood** Students make plywood from thin strips of wood and glue. Students compare the breakable strength of a craft stick to that of their homemade plywood.	**Introduction** 10 minutes **Center** 20 minutes **Notebook** 20 minutes **Reading** 20 minutes	How is plywood made?

Content	Writing/Reading	Assessment
• Wood has many observable properties. • Sanding can change the shape of wood. • Sawdust and wood shavings are tiny pieces of wood.	**Science Notebook Entry** Draw or write words to answer the focus question.	**Embedded Assessment** Teacher observation
• Wood has many observable properties, including whether it floats or sinks in water. Some pieces of wood float on top; some sink to the bottom. • Wood that is waterlogged sinks. • A mixture consists of two materials mixed together. • Water dries up when left in an open container.	**Science Notebook Entry** Draw or write words to answer the focus questions.	**Embedded Assessment** Teacher observation
• Some objects occur in nature; others are made by people. • Sawdust can be changed into particleboard. • Water left in an open cup dries up and goes into the air (evaporates). Water left in a closed cup does not evaporate.	**Science Notebook Entry** Draw or write words to answer the focus question.	**Embedded Assessment** Teacher observation
• Some objects occur in nature; others are made by people. • Wood pieces can be changed into plywood. • Gluing (laminating) thin sheets of wood together produces stronger wood that is hard to break.	**Science Notebook Entry** Draw or write words to answer the focus question. **Science Resources Book** "Are You a Scientist?"	**Embedded Assessment** Teacher observation

BACKGROUND *for the Teacher*

This investigation is about changing wood, in particular, the ways people change wood. Humans first change tree trunks into lumber and related manageable chunks, and then they change the lumber into products.

Wood production usually begins with the felling of a living tree rather than a dead one. Living trees are more abundant, the wood is in better condition, and they are easier to handle and less likely to split or break during harvest and transportation. Logs are hurried to the lumber mill and kept wet to prevent cracking and splitting until they are sawed. Freshly milled lumber is called green lumber, not because of its color, but because it is still filled with moisture and sap. Green lumber is usually air dried (a lengthy process) or kiln dried (a quick process) before it is used.

How Can You Change the Shape of Wood?

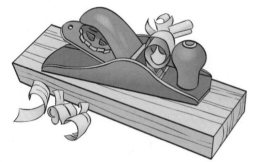

Changing the **shape** of wood has been a human enterprise since time immemorial. Doubtless the first change was accomplished by simple brute force—the breaking of a suitable length of tree limb for a weapon or the disassembly of fallen timber for firewood. When our predecessors acquired technology, stone axes ushered in the art of **woodworking**. Tools made it possible to hew wood as well as to break it. Boats, crafted weapons, housing, and a host of other civil trappings resulted. Metal tools increased the efficiency and versatility of woodworking. Today, humans have a daunting array of hand tools and power tools for transforming wood: axes, adzes, knives, bits, planes, sanders, rasps, saws, chisels, machetes, mills, and chain saws. We can transform the mightiest cedar into lumber in a wink with powerful machinery, and then, in the hands of a craftsperson, again transform that cedar into a light, tough, and exquisitely beautiful traditional canoe.

How Are Sawdust and Shavings the Same? How Are Sawdust and Shavings Different?

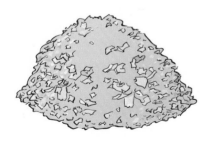

The industrial processing of wood is done in large part by hewing—cutting away bits of wood to change its size and shape. Sawing removes a strip of wood the width of the saw blade as the teeth gouge and scrape away the wood in its path. The removed wood, called **sawdust**, is a waste product. So, too, are the by-products of planing (flattening), jointing (shaping), and drilling lumber. For years, the sawdust and **shavings** were burned.

Sawdust and shavings are names for small pieces of wood. They are both made of the same material and differ only in size. The smallest of all wood particles are those created when wood is **sanded**. Coarse sandpapers produce gritty bits of wood, but the finest grades of **sandpaper** produce true dust. The particles are so small, they are impossible to feel. When sawdust or shavings are put into water, they float, but only briefly, because the small size of the particles allows the water to fill all the air spaces very quickly. The substances from which wood is made are denser than water, so once the air in the wood particles has been displaced by water, the piece of wood is **waterlogged**, rendering it denser than water, so down it drifts. Students will observe the flecks of wood to be almost neutrally buoyant, suspended in water, creating little aquatic environments in their cups, reminiscent of a wintery scene in the ever-popular snow globe.

How Is Particleboard Made?

In the last century, technology made it possible to transform wood into even more versatile materials and products. One material that is technologically derived is **particleboard**. This material was invented to take advantage of abundant waste wood fiber. As the technology and demand for the product advanced, it became feasible to create particles specially for the manufacture of particleboard. Today, a carefully graded batch of particles is sprayed liberally with an adhesive resin. The "liquid" wood slurry enters a carefully engineered space where the particles are lifted with jets of air. The system allows the particles to fall to the surface in an interesting way. The smallest particles precipitate first, followed by successively larger particles, and then the process reverses so the last particles are again the very small ones. The result is that the top and bottom surfaces are composed of the finest particles, giving the finished sheet a very smooth surface. The raw sheet is then pressed and heated to finish the product. Particleboard is used to enclose spaces, as subflooring in building construction, and for the sheet work in cabinetry and furniture that is destined to be covered with an attractive presentation surface, such as wood veneer or a plasticized finish.

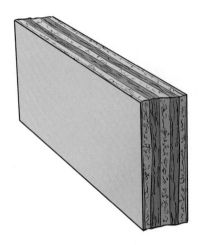

How Is Plywood Made?

Plywood is made from thin sheets of wood **glued** together, with alternate **layers** of grain running in perpendicular directions. This construction produces a tremendously **strong**, large surface of wood, much more suitable for walls and floors than is conventional lumber. Conventional lumber is very strong if you try to **break** it across the grain, but not nearly so strong if broken along the grain. Alternating the grain direction in plywood makes it strong in all directions.

Plywood is made by first transforming a log into veneer on a machine called a peeler. A rotating log is thrust against a long blade, which slowly advances toward the center of the log. The wood is transformed into one continuous sheet a few millimeters thick. The sheet is then cut into uniform rectangles, glued, and stacked as described above. The sheets are heated under pressure to cure the glue.

Wood is not a highly recyclable material. Some wood is reclaimed from old buildings that are being razed; it is remilled and used in construction a second time, but that's about it. Any wood that is pitched out into the environment at the end of its useful life will soon be recycled by natural processes. Even though most of the life-forms with which we are most familiar derive no nutritional value from cellulose, a lot of other organisms (such as fungi and bacteria) find cellulose quite to their liking. Wood is a biodegradable, renewable resource. It is a natural material that we have thoroughly integrated into human culture.

New Word — Say it • See it • Hear it • Write it

Break
Change
Cornstarch
Dries up
Evaporate
Glue
Laminated
Laminating
Layer
Mix
Mixture
Sand
Sandpaper
Sawdust
Screen
Shape
Shavings
Strong
Waterlogged
Woodworker

TEACHING CHILDREN *about* *Changing Wood*

Wood is excellent for fabrication because not only can it be transformed into many shapes and sizes, but it also combines effectively with other materials to make new materials. That's what this investigation centers on—change and transformation.

The simple process of using sandpaper on wood allows students to see transformations. A relatively large piece of wood is changed into a smaller piece, and the piece has a different shape. At the same time, part of the wood is reduced to dust. The changed piece of wood and the dust add up to the original piece, but things are different, and students know why they are different—change was imposed on the materials by the students themselves.

The **conceptual flow** for this investigation starts in Part 1 with students sanding a piece of wood. Their **actions produce change**. The starting piece of wood is a **new shape**, and there is now a little pile of dust.

In Part 2, students work with two sizes of wood **particles** identified as sawdust and shavings. The particles are mixed and stirred, showing that smaller particles go to the bottom of the cup. When the particles are dumped into water, they **float** briefly and then **sink**. This **behavior** is contrary to the floating of wood blocks. Properties of wood change when wood soaks in water: the wood gets **waterlogged** and sinks. When the wood particles are separated from the water and laid out on paper, they change from wet to **dry** when the water "dries up" (evaporates).

In Part 3, students make **particleboard** by mixing wood particles with an **adhesive**—cornstarch glue (matrix). After a period of drying, the wood particles have changed into a robust, solid, **new material**.

In Part 4, students make **plywood** by gluing (laminating) three thin pieces of wood to make a **stronger** product (**laminated**).

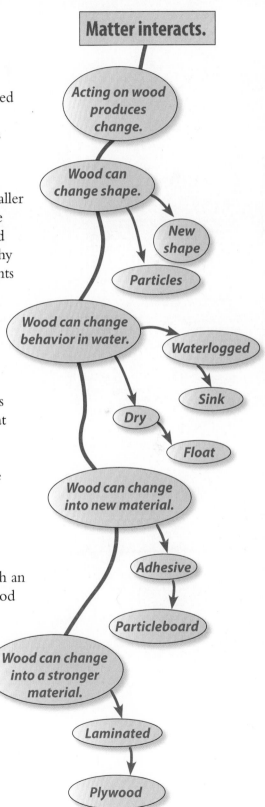

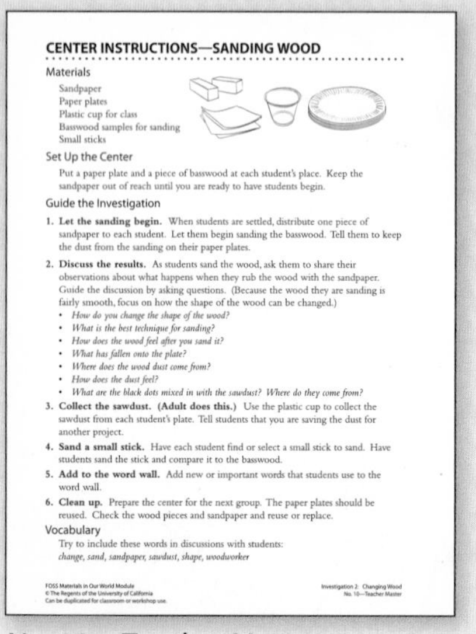

CENTER INSTRUCTIONS—SANDING WOOD

Materials

Sandpaper
Paper plates
Plastic cup for class
Basswood samples for sanding
Small sticks

Set Up the Center

Put a paper plate and a piece of basswood at each student's place. Keep the sandpaper out of reach until you are ready to have students begin.

Guide the Investigation

1. **Let the sanding begin.** When students are settled, distribute one piece of sandpaper to each student. Let them begin sanding the basswood. Tell them to keep the dust from the sanding on their paper plates.
2. **Discuss the results.** As students sand the wood, ask them to share their observations about what happens when they rub the wood with the sandpaper. Guide the discussion by asking questions. (Because the wood they are sanding is fairly smooth, focus on how the shape of the wood can be changed.)
   - *How do you change the shape of the wood?*
   - *What is the best technique for sanding?*
   - *How does the wood feel after you sand it?*
   - *What has fallen onto the plate?*
   - *Where does the wood dust come from?*
   - *How does the dust feel?*
   - *What are the black dots mixed in with the sawdust? Where do they come from?*
3. **Collect the sawdust. (Adult does this.)** Use the plastic cup to collect the sawdust from each student's plate. Tell students that you are saving the dust for another project.
4. **Sand a small stick.** Have each student find or select a small stick to sand. Have students sand the stick and compare it to the basswood.
5. **Add to the word wall.** Add new or important words that students use to the word wall.
6. **Clean up.** Prepare the center for the next group. The paper plates should be reused. Check the wood pieces and sandpaper and reuse or replace.

Vocabulary

Try to include these words in discussions with students:
change, sand, sandpaper, sawdust, shape, woodworker

FOSS Materials in Our World Module
© The Regents of the University of California
Can be duplicated for classroom or workshop use.

Investigation 2: Changing Wood
No. 10—Teacher Master

No. 10—Teacher Master

MATERIALS *for*

Part 1: *Sanding Wood*

For each student

- 1 Piece of sandpaper (See Step 6 of Getting Ready.)
- 1 Basswood sample for sanding (See Step 7 of Getting Ready.)
- 1 Paper plate
- 1 Collecting bag (See Step 8 of Getting Ready.) ★
- 1 Stick, about 20 cm (8") long × 1 cm (0.5") thick (See Step 4 of Getting Ready.) ★

For the class

- 1 Plastic cup
- ❑ 1 Teacher master 2, *Focus Questions A*
- ❑ 1 Teacher master 10, *Center Instructions—Sanding Wood*
- • Safety goggles (for sanding) ★

For assessment

- ❑ • *Assessment Checklist*

★ Supplied by the teacher. ❑ Use the duplication master to make copies.

GETTING READY *for*
Part 1: *Sanding Wood*

1. Schedule the investigation

If you have a suitable outdoor location, consider doing the sanding outdoors with the whole class at one time. It will take about 30 minutes. If you do this indoors, the part requires 20 minutes at the center for each group of six to ten students.

Plan 5 minutes to introduce and 20 minutes for recording in notebooks and to wrap up.

2. Preview Part 1

Students add to their knowledge of the properties of wood and learn how to use those properties to change wood. Students use sandpaper to change the shape of basswood and a stick. The focus question is **How can you change the shape of wood?**

3. Consider safety

If a student has respiratory problems, such as asthma, use a breathing mask, handkerchief, or bandanna to cover nose and mouth. For these students, check with parents or guardians to make sure that sanding wood is OK. Check your district safety guidelines for the use of safety goggles while sanding, and provide them as necessary.

▶ SAFETY NOTE

Be aware of students with respiratory issues, such as asthma, and provide adequate protection as necessary.

4. Select your outdoor site

This activity is best done on a calm day outdoors. Search for a level area of ground for the class to sit on while they are sanding. The area should be sheltered from the wind and near a location where students can find sticks.

Each student will need one small stick or twig. If there aren't enough sticks in the location, provide sticks from elsewhere, one for each student.

5. Check the site

It is always a good idea to check the outdoor site on the morning of an outdoor activity. Check for any distracting items both in the area where students will sand as well as where they will find sticks.

6. Tear the sandpaper

If you are the first person to use the kit, you will need to tear the sandpaper before starting this activity. (Sandpaper will dull your scissors if you try to cut it.) First, fold each sheet of sandpaper in thirds. Fold it sharply in both directions so you will be able to tear it cleanly into three strips along the fold lines. Fold and tear each strip into three equal parts. The sandpaper is ultimately a consumable item, but should be reused as much as possible.

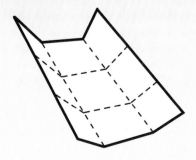

> **NOTE**
>
> Refer to the *Taking FOSS Outdoors* chapter for details on managing materials outdoors. A student collecting bag can be fashioned from a zip bag and string.

7. **Prepare basswood samples**

   The kit has 72 basswood samples. One package has 36 samples that are permanent equipment; a second package has 36 consumable pieces for sanding. Make sure the packing bags are clearly marked. Use the *sanding* samples for this investigation.

8. **Prepare materials for going outdoors**

   Set the materials at a materials station in the classroom. Each student will use a collecting bag to pick up and carry a paper plate, basswood sample, and piece of sandpaper to the outdoor sanding site. Remember to save the sandpaper and paper plates—they are to be reused.

9. **Set up the center (indoors optional)**

   If you decide to do this activity indoors, put a paper plate and a piece of basswood at each student's place. Keep the sandpaper out of the reach of students until you are ready for them to begin. Remember to save the sandpaper and paper plates—they are to be reused.

10. **Plan assessment**

    Three objectives can be assessed at any time during any part of this investigation.

    What to Look For

    - *Students ask questions.*

    - *Students communicate observations orally.*

    - *Students use new vocabulary.*

    Here are specific objectives to observe in this part.

    - *Wood shape and texture can be changed by sanding.*

    - *Students use tools appropriately.*

    Focus on a few students each session. Record the date and a + or − on the *Assessment Checklist.*

GUIDING *the Investigation*
Part 1: *Sanding Wood*

1. **Introduce the investigation**

   Call students to the rug. Hold up a piece of **sandpaper** and ask if they know what it is. Confirm that it is sandpaper—paper with **sand** glued on it. Tell students that **woodworkers** use sandpaper to **change** the **shape** of wood, take off splinters, and make wood surfaces smooth.

2. **Review basswood**

   Show students a basswood sample. Ask them to look at the piece of wood and recall its name. Tell them that today their challenge will be to change the shape of the wood.

3. **Focus question: How can you change the shape of wood?**

   Write the focus question on the chart, and read it together.

   ➤ *How can you change the shape of wood?*

   Ask students for ideas on how to use the sandpaper to change the shape of the wood. Listen to their ideas, and then model how to rub the sandpaper on the basswood.

4. **Go outdoors**

   Quickly review the procedures and behavior expectations for outdoor learning.

   Before going outdoors, have students pass through the materials station and get a collecting bag. In the bag, they should put one plate, one sample of basswood, and one piece of sandpaper.

   Once outside, have students sit on the ground with their plates in front of them and begin sanding. Instruct students to try to keep the dust from the sanding on their paper plates.

   NOTE: Caution students not to rub their eyes when they are sanding. Provide safety goggles based on your district guidelines.

   If you choose to do this indoors at a center, send six to ten students to the center. Distribute one small piece of sandpaper to each student. Let students begin sanding the basswood. Tell them to keep the dust from the sanding on their paper plates.

FOCUS QUESTION

How can you change the shape of wood?

Say it · See it · Hear it · Write it — **New Word**

Materials for Step 4
- *Collecting bags*
- *Paper plates*
- *Basswood samples*
- *Sandpaper pieces*
- *Plastic cup*
- *Safety goggles*

▶ **SAFETY NOTE**

Be aware of students with respiratory issues, such as asthma, and provide adequate protection as necessary.

5. **Discuss results**

As students sand the wood, ask them to share their observations about what happens when they rub the wood with the sandpaper. Because the wood they are sanding is already fairly smooth, focus on how the shape of the wood can be changed. Guide the discussion by asking questions.

➤ *How do you change the shape of the wood?*

➤ *What is the best technique for sanding?*

➤ *How does the wood feel after you sand it?*

➤ *What has fallen onto the plate?*

➤ *Where does the sawdust wood come from?*

Introduce the word *sawdust*. Tell students that sanding wood makes **sawdust**.

➤ *How does the sawdust feel?*

➤ *What are the black dots mixed in with the sawdust? Where do they come from?* [Pieces of sand from the sandpaper.]

6. **Collect sawdust**

Use the plastic cup to collect the sawdust from each student's plate. Students can later tape bits of the sawdust into their notebooks.

7. **Find a stick to sand**

Have the class put their basswood, sandpaper, and plates back in their collecting bags. Walk the class to the area where they will search for their sticks. Hold up a stick that is about the right size, 20 cm long (8") and at least 1 cm thick (about 0.5"). Challenge each student to find one good stick to sand. Then return to the sanding area.

8. **Compare the stick to the basswood**

As they sand, ask students to compare sanding the stick to sanding the basswood. They might find that the stick is easier to hold and easier to sand.

9. **Return to class**

Have students bring their sticks inside and share them at the rug.

If you have been doing this activity inside, prepare the center for the next group. The paper plates should be reused. The wood pieces and sandpaper can also be reused, but you will have to check the condition of both and replace them if necessary.

10. Review vocabulary

Review key vocabulary, and add it to the word wall. Here's a suggested cloze review. Students answer chorally.

change
sand
sandpaper
sawdust
shape
woodworker

➤ *Paper with sand glued on it is called _____ .*

S: Sandpaper.

➤ *We can change the shape and smooth the wood. We use sandpaper to _____ .*

S: Sand.

➤ *A person who works with wood is called a _____ .*

S: Woodworker.

➤ *When we rub the wood with sandpaper, the dust that comes off is called _____ .*

S: Sawdust.

Discuss these additional words, and write them on the word wall.

- Change
- Shape

11. Answer the focus question

Restate the focus question written on the chart, and read it with students.

➤ *How can you change the shape of wood?*

Tell students you have a strip of paper with the question written on it. Review how to glue the strip into the notebook.

Ask students to answer the focus question in drawings and/or words. Students might draw the shape of the wood before sanding and after sanding. Students might also tape bits of sawdust into their notebooks.

FOCUS CHART

How can you change the shape of wood?

Use _____ to change the shape of wood.

WRAP-UP/WARM-UP

12. Share notebook entries

Conclude Part 1 or start Part 2 by having students share notebook entries. Ask students to open their science notebooks to the most recent entry. Read the focus question together as a class.

➤ *How can you change the shape of wood?*

Ask students to pair up with a partner to

- share their answers to the focus question;
- explain their drawings.

MATERIALS *for*

Part 2: *Sawdust and Shavings*

For each student at the center

- 1   Plastic cup
- 1   Container, 1/2 L
- 1   Craft stick
- 1   Paper plate
- 1   Screen

For the class

- 1   Plastic spoon ★
- 1   Bag of sawdust (smaller particles)
- 1   Bag of wood shavings (larger particles)
- 2   Paper plates
- 6   Plastic cups
- 3   Lids for plastic cups
- •   Water ★
- 1   Pitcher or empty 2 L soft-drink bottle ★
- •   Paper towels ★
- •   Newspaper ★
- 1   Saw (optional) ★
- 1   Plane (optional) ★
- •   Safety goggles ★
- ❏ 1   Teacher master 11, *Center Instructions—Sawdust and Shavings*

For assessment

- •   *Assessment Checklist*

★ Supplied by the teacher.   ❏ Use the duplication master to make copies.

No. 11—Teacher Master

GETTING READY *for*
Part 2: *Sawdust and Shavings*

1. **Schedule the investigation**
   This part requires 20 minutes at the center for each group of six to ten students. Plan 5 minutes to introduce and 20 minutes for recording in notebooks and to wrap up. Plan on 10 minutes a few days later to observe the water in the closed and open cups.

2. **Preview Part 2**
   Students compare sawdust and shavings. Students find out what happens to sawdust and shavings when they mix the two with water and then separate out the shavings. Students spread out wet sawdust on paper plates and put some in a closed container. The focus questions are **How are sawdust and shavings the same? How are sawdust and shavings different?**

3. **Consider safety**

   If a student has respiratory problems, such as asthma, use a breathing mask, handkerchief, or bandanna to cover nose and mouth or have that student use only wood shavings. Check your district safety guidelines for the use of safety goggles, and provide them as necessary.

4. **Prepare for the introduction**
   Put a heaping spoonful of sawdust on one paper plate and a heaping spoonful of wood shavings on another paper plate.

5. **Obtain a saw and plane (optional)**
   If possible, get a saw and plane to show students how to produce sawdust and shavings. Talk to your school custodian or a shop teacher, or invite a woodworker to bring tools and give a demonstration. Use appropriate safety practices.

6. **Prepare sawdust and wood shavings**
   Put one heaping spoonful of sawdust in a plastic cup and one heaping spoonful of wood shavings in a 1/2-liter (L) container for each student.

7. **Set up the center**
   Put all the materials on the table or a counter close by for easy distribution to students.

8. **Plan for cleanup**
   At the end of this part, remove most of the water from the sawdust mixture, and spread the solid material out to dry on newspaper.

9. **Plan assessment**

Here are specific objectives to observe in this part.

- *Two materials mixed together make a mixture.*

- *Some materials float in water; others sink.*

- *Some materials absorb water.*

- *Students compare properties of wood pieces.*

- *Water dries up when left in an open container.*

Focus on a few students each session. Record the date and a + or − on the *Assessment Checklist*.

GUIDING *the Investigation*
Part 2: *Sawdust and Shavings*

1. **Introduce sawdust and wood shavings**
   Call students to the rug. Show them the sawdust and shavings on the paper plates. Ask students how the materials are the same. [They are both small pieces of wood.] Ask students how they are different. [The sawdust pieces are smaller and a different color.]

   If you brought in a saw and plane, or invited a guest to bring in his or her tools, show students the tools, and discuss how they are used and which tool produces sawdust and which produces shavings.

2. **Focus questions: How are sawdust and shavings the same? How are sawdust and shavings different?**
   Write the focus questions on the chart, and read them together.

   ➤ *How are sawdust and shavings the same?*

   ➤ *How are sawdust and shavings different?*

3. **Explore the sawdust**
   Send six to ten students to the center. Give each student a cup containing sawdust and a craft stick. Let students explore the sawdust with their craft sticks for a minute or two.

   Be aware of students with respiratory issues such as asthma and provide adequate protection as necessary.

Materials for Steps 3–4
- *Cups of sawdust*
- *Containers of wood shavings*
- *Craft sticks*
- *Safety goggles*

4. **Explore the wood shavings**
   Give each student a 1/2 L container with wood shavings. Give students 2–3 minutes to explore this new wood material with their craft sticks.

5. **Combine the two materials**
   Ask students to pour the sawdust into the wood shavings container. Have them use sticks to stir the mixture in the container. Ask,

   ➤ *Where is the sawdust now?* [Most of the sawdust will end up on the bottom.]

6. **Introduce *mixture***
   Tell students,

   *You made a **mixture** of sawdust and **shavings**. When you put two or more materials together and stir them up, you have a mixture.*

   Have students say "mixture." Write it on the board. Clap the two syllables of *mixture* together.

Materials for Step 7
- *Cups of water*

7. Drop sawdust and shavings in water

Half fill an empty cup with water for each student. Ask what students think will happen if they sprinkle the mixture of sawdust and wood shavings on the water. Let students sprinkle their sawdust and shavings mixture on the water, a little bit at a time, without stirring. Encourage students to look through the side of the cup. After 2 minutes, let students stir their cups.

8. Discuss the wood in the water

Ask students to report what they observed.

> ➤ *What material is sawdust made of?* [Wood. Sawdust is small pieces of wood.]

> ➤ *What material are shavings made of?* [Wood. Shavings are small pieces of wood.]

> ➤ *Where are the small pieces of wood in the cup of water?* [Some floated, some floated for a while and sank, some sank to the bottom right away.]

> ➤ *Are the small pieces of wood above the surface, below the surface, or on the bottom?*

TEACHING NOTE

This question is asking students to describe what happens to the pieces of wood (both sawdust and shavings) when they are put in water. Students aren't expected to describe differences in the behavior of sawdust and shavings.

Say it
Hear it
See it
Write it
New Word

---

9. Introduce *waterlogged*

Tell students,

When water soaks into wood, we say the wood is **waterlogged**. *Waterlogged wood sinks to the bottom of the cup of water.*

Have students say "waterlogged." Write it on the board. Clap the three syllables of *waterlogged* together.

Materials for Step 10
- *Screens*

E L   N O T E

Make sure that students understand what the word separate means.

---

10. Strain out the wood pieces

Ask students how they could separate the water and wood pieces. After discussing their ideas, show them a tool made for the job—a **screen**. Have students place their screens on top of the 1/2 L containers and then pour the water, sawdust, and shavings mixtures from the cups through the screens, catching the water in the containers.

11. Investigate wet wood pieces

Have students dump the soggy sawdust and shavings caught in the screen onto a paper plate. Allow time for students to explore the wet shavings.

12. Discuss the wet wood

Ask students,

➤ *How did the sawdust and wood shavings change after you added water?* [They are wet, stick together a little bit, and may be darker.]

Put a cup of dry shavings and sawdust on the table. Ask,

➤ *What could you do to get the wet sawdust and shavings to look like they did before we added water?* [Let them dry in the air.]

Show students how to spread the wet sawdust and shavings on a paper plate to dry. Students will come back tomorrow and see what happened.

13. Consider the remaining water in the cup

Ask students,

➤ *There is water left in the cup. What do you think will happen to the water if we set the cup on the table until tomorrow? Will the water dry up overnight or in a few days?*

Suggest that students leave one cup with just a little bit of water on the bottom and revisit it to see what happens over time. For comparison, put the same amount of water in a second cup and cover it with a lid. Leave the two cups near the paper plates with the wet shavings.

14. Clean up

Get the center ready for the next group. Wipe out the cups so they are dry. Put a spoonful of shavings in the containers and a spoonful of sawdust in the cups.

NOTE: Allow the shavings and sawdust to dry on the paper plates. Save the dried sawdust and shavings for Part 3.

15. Review vocabulary

Review key vocabulary added to the word wall. Here's a suggested cloze review. Students answer chorally.

➤ *Thin pieces of wood that have been shaved off a larger piece of wood are called* _____ .

S: Shavings.

Materials for Step 11
- *Paper plates*
- *Lids for cups*
- *Paper towels*
- *Newspaper*

 mixture
screen
shavings
waterlogged

➤ *When water soaks into wood, we say the wood is ____ .*

S: Waterlogged.

➤ *Things that are mixed together are called a ____ .*

S: Mixture.

➤ *To separate the water from the wood pieces, we used a ____ .*

S: Screen.

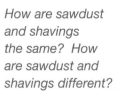

FOCUS CHART

How are sawdust and shavings the same? How are sawdust and shavings different?

Both sawdust and shavings are ____ .

Sawdust is ____ .

16. Answer the focus questions

Restate the focus questions.

➤ *How are sawdust and shavings the same?*

➤ *How are sawdust and shavings different?*

Tell students you have a strip of paper with the questions written on it. Review how to glue the strip into the notebook.

Ask students to answer the focus questions in drawings and/or words.

WRAP-UP/WARM-UP

17. Share notebook entries

Conclude Part 2 or start Part 3 by having students share notebook entries. Ask students to open their science notebooks to the most recent entry. Read the focus questions together as a class.

➤ *How are sawdust and shavings the same?*

➤ *How are sawdust and shavings different?*

Ask students to pair up with a partner to

- share their answers to the focus questions;
- explain their drawings.

B R E A K P O I N T

18. Make observations after several days

After several days, have students observe the sawdust and shavings and the cups of water left out with the shavings. Ask,

➤ *What happened to the sawdust and shavings?* [They got dry.]

➤ *How are the two cups of water different?* [The open one is dry; the closed one still has water in it.]

➤ *Where did the water go that was in the shavings and the open cup?* [It dried up; it went into the air.]

TEACHING NOTE

*See the **Home/School Connection** for Investigation 2 at the end of the Interdisciplinary Extensions section. This is a good time to send it home with students.*

MATERIALS *for*

Part 3: *Making Sawdust Wood*

For each student at the center

- 1   Plastic cup (with sawdust and shavings)
- 1   Craft stick
- 1   Paper plate
- 1   Particleboard sample

For the class

- 2   Pine samples
- 2   Particleboard samples
- •   Sawdust and shavings (dried, from Part 2)
- 1   Pair of evaporation cups, one with lid (from Part 2)
- 2   Containers, 1/2 L
- 2   Plastic spoons ★
- 1–2   Boxes of cornstarch ★
- •   Water ★
- 1   Pencil or marker ★
- •   Scratch paper ★
- •   Paper towels ★
- 1   Saucepan ★
- 1   Long-handled spoon ★
- 1   Container, large zip bag, or jar ★
- 1   Poster, *Particleboard Production*
- ❏   1   Teacher master 12, *Center Instructions—Making Sawdust Wood*

For assessment

- •   *Assessment Checklist*

★ Supplied by the teacher.   ❏ Use the duplication master to make copies.

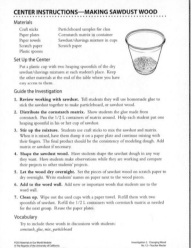

No. 12—Teacher Master

GETTING READY *for*
Part 3: *Making Sawdust Wood*

1. **Schedule the investigation**

   This part requires 20 minutes at the center for each group of six to ten students and 10 minutes to introduce. Plan on 20 minutes for recording in notebooks and to wrap up. Three days later, plan about 10 minutes for discussion with the whole class after the projects have dried.

2. **Preview Part 3**

   Students simulate the making of particleboard by using sawdust and a cornstarch matrix. Students compare their particleboard with the samples from the kit. The focus question is **How is particleboard made?**

3. **Make cornstarch matrix**

   A matrix holds the sawdust together and forms a dough from which a wood sample can be shaped. Make this matrix the *day before* you will be using it at the center. The following recipe makes enough for 24 students.

   a. Stir constantly while adding one box of cornstarch gradually to 3.5 cups of cold water in a saucepan.

   b. Heat the mixture over medium heat, continuing to stir. Keep stirring until about half the matrix has thickened to the consistency of soft mashed potatoes. The rest will be soupy. This should take about 5–10 minutes.

   NOTE: Don't cook the cornstarch matrix until it gets thick—when it cools, it will be too thick to use. If it does seem too thick when you're ready to use it, add water, a little at a time.

   c. Remove the matrix from the heat, and stir it until it is consistent and cool. It may get as thick as pudding. Store it in a covered container, zip bag, or jar.

   Keep the matrix refrigerated; it will keep for up to 1 week. If the mixture gets too thick to pour when you're ready to use it, stir in a little water to thin it.

4. **Prepare the sawdust**

   Put two heaping spoonfuls of *dry* sawdust and shavings in a plastic cup for each student.

5. **Set up the center**

   Put a plastic cup with sawdust and shavings, a craft stick, a paper plate, and a particleboard sample at each student's place. Keep the other materials (spoons, cornstarch matrix, paper towels, and wood samples) where you will have easy access.

6. **Plan assessment**

   Here are specific objectives to observe in this part.

   - *Wood has observable properties.*

   - *Common materials can be made into new materials. Sawdust can be recycled to make wood.*

   - *Students compare properties of wood.*

   - *Water dries up when left in an open container.*

   Focus on a few students each session. Record the date and a + or − on the *Assessment Checklist*.

How is particleboard made?

Materials for Step 3
- *Evaporation cups (one with lid)*

Materials for Step 4
- *Pine samples*
- *Particleboard samples*
- **Particleboard Production** *poster*

"Sawdust wood" is the descriptive name students sometimes give to particleboard.

GUIDING *the Investigation*
Part 3: *Making Sawdust Wood*

1. **Recall the sawdust and shavings**

   Call students to the rug. Have students recall their experiences with sawdust and shavings in Part 2. Remind them that at the end of the session, they spread the wet sawdust and shavings on paper plates.

   Retrieve the dried sawdust and shavings and put them in a cup. Pass it around so students can feel it and see that it is dry. Ask,

   ➤ *What happened to the water that was on the sawdust and shavings on the paper? Where did it go?* [It **dried up** and went into the air.]

   Confirm that when the wet sawdust was open to the air, the water dried up.

2. **Introduce** *evaporate*

   Tell students that another word that describes water drying up is *evaporate*. Explain that when water evaporates, it goes into the air.

   Have students say "evaporate." Write it on the board. Together, clap the four syllables of *evaporate*.

3. **Check cups of water**

   Remind students that they also had two cups of water, one with a lid on it and one without. Hold up the two cups, and ask students to observe the water in the cups. Ask if there is still the same amount of water in the two cups. Depending on the number of days that has elapsed, students may or may not observe a difference. This will be an ongoing investigation for students to observe.

4. **Observe the particleboard samples**

   Hold up the samples of pine and particleboard. Remind the class that one kind of wood came right from the tree and that the other started from a tree, but was changed and processed by people. Ask students which wood came right from the tree.

   Review the poster that shows how particleboard is made. Tell students that they will each make a piece of particleboard when they work at the center.

5. **Focus question: How is particleboard made?**

   Write the focus question on the chart, and read it together.

   ➤ *How is particleboard made?*

   Tell students that today they will use homemade **glue** to stick the sawdust together to make **particleboard**—or sawdust wood.

6. Distribute the cornstarch mixture

Send six to ten students to the center. Show them the two 1/2 L containers of **cornstarch** matrix. Tell them that it is a kind of glue made from a mixture of cornstarch and water. You might show them a package of cornstarch.

Put a plastic spoon in each of the 1/2 L containers of cornstarch matrix and pass the containers around the group. Have each student put one heaping spoonful of the matrix in his or her cup of sawdust.

7. Stir up the mixture

Have students use their craft sticks to **mix** the sawdust and matrix. When the mixture is fairly uniform, have students dump it onto their paper plates and use their fingers to continue mixing. The final product should be the consistency of modeling dough.

Monitor students' progress and add additional matrix or sawdust if necessary. The sawdust wood should hold together well. If it crumbles or cracks, add more matrix.

8. Shape the sawdust wood

When students have a ball of sawdust dough, let them shape the wood in any way they want. Many will want to shape it like the particleboard sample, but others may choose to make other shapes. Have students make observations while they are working. Compare their projects to other students' projects.

9. Add vocabulary to the word wall

As students offer their observations, add any new or important vocabulary to the word wall.

10. Let the wood dry overnight

Set the newly formed pieces of sawdust wood on scratch paper to dry overnight. Write students' names on the paper next to each of their wood pieces.

11. Clean up

Use paper towels to wipe out the cups. Refill the cups with two spoonfuls of sawdust and the 1/2 L containers with cornstarch matrix as needed for the next group. Reuse the paper plates.

When all the groups have completed the activity, wash and dry the 1/2 L containers, plastic cups, spoons, and craft sticks before putting them back in the kit. Keep the paper plates that are still in usable condition, and recycle the plates that can't be reused.

Materials for Steps 6–11
- *Containers of cornstarch matrix*
- *Plastic spoons*
- *Cups of sawdust*
- *Craft sticks*
- *Paper plates*
- *Scratch paper*
- *Particleboard samples*
- *Paper towels*

Say it
New Word
See it
Hear it
Write it

cornstarch
dries up
evpoarate
glue
mix
particleboard

12. Review vocabulary

Review key vocabulary added to the word wall. Here's a suggested cloze review. Students answer chorally.

➤ *Water left out in the open air _____ .*

S: Evaporates.

➤ *What kind of wood did we make today?*

S: Particleboard.

➤ *We used a cornstarch and water mixture for our _____ .*

S: Glue.

➤ *We used craft sticks to do this to the sawdust and matrix.*

S: Mix.

B R E A K P O I N T

Materials for Step 13
- *Dried sawdust wood*
- *Particleboard samples*

FOCUS CHART

How is particleboard made?

We _____ to make particleboard.

13. Answer the focus question

After the sawdust-wood pieces have dried, have students retrieve their pieces and compare them to the particleboard samples in the kit. Guide a discussion by asking if sawdust wood grows on sawdust-wood trees.

➤ *How is particleboard made?*

Tell students you have a strip of paper with the question written on it. Review how to glue the strip into the notebook.

Ask students to answer the focus question in drawings and/or words.

You might discuss these additional questions with students.

➤ *How is your sawdust wood like the particleboard?*

➤ *Why do you think people make particleboard?* [To recycle waste wood, such as sawdust and shavings.]

WRAP-UP/WARM-UP

14. Share notebook entries

Conclude Part 3 or start Part 4 by having students share notebook entries. Ask students to open their science notebooks to the most recent entry. Read the focus question together as a class.

➤ *How is particleboard made?*

Ask students to pair up with a partner to

- share their answers to the focus question;
- explain their drawings.

MATERIALS *for*
Part 4: *Making Sandwich Wood*

For each student

- 3   Thin wood pieces
- 1   Sheet of newspaper ★
- 1   *FOSS Science Resources: Materials in Our World*
  - •   "Are You a Scientist?"

For the class

- 1   Plywood sample
- 5   Pieces of thin plywood pieces, 1/4"
- 2–3   Craft sticks
- 1   Poster, *Plywood Production*
- 3   Thin wood pieces
- 1   Sheet of newspaper ★
- •   White glue ★
- 1   Pencil or marker ★
- •   Paper towels ★
- ❑   1   Teacher master 13, *Center Instructions—Making Sandwich Wood*
- 1   Big book, *FOSS Science Resources: Materials in Our World*

For assessment

- •   *Assessment Checklist*

★ Supplied by the teacher.   ❑ Use the duplication master to make copies.

No. 13—Teacher Master

CENTER INSTRUCTIONS—MAKING SANDWICH WOOD

Materials

Newspaper mats   Thin wood pieces
White glue   Pencil
Paper towels   Craft sticks
1/4" plywood samples

Set Up the Center

Put a sheet of folded newspaper at each student's place. Pile the thin wood pieces in the center of the table where students can easily reach them. Have the white glue ready for distribution when needed. Have the 1/4" plywood samples ready for students to compare to what they make.

Guide the Investigation

1. **Make sandwich wood.** Have each student choose three thin wood pieces. Go around the group and squeeze a quarter-size pool of glue onto the far side of each student's newspaper mat. Let students begin gluing the pieces together.

   NOTE: Make sure students put glue on both surfaces they are gluing together, spread the glue over the entire surface of each piece, and hold the pieces together for a count of 20 after each piece of wood is added.

2. **Add to the word wall.** Add new or important words that students use to the word wall.

3. **Make comparisons.** Put the 1/4" plywood samples around the table. Ask students to compare the plywood samples to the plywood they made.

4. **Clean up.** Find a location where the wood can dry overnight. Label students' pieces of plywood by writing their names on the wood sandwich. Replace the newspaper mats for the next group of students, or use the same sheets turned inside out if they are not covered with too much glue.

Vocabulary

Try to include these words in discussions with students: *break, laminated, layer, plywood, strong*

FOSS Materials in Our World Module
© The Regents of the University of California
Can be duplicated for classroom or workshop use.

Investigation 2: Changing Wood
No. 13—Teacher Master

GETTING READY *for*
Part 4: *Making Sandwich Wood*

1. **Schedule the investigation**

   This part requires 20 minutes at the center for each group of six to ten students and 10 minutes to introduce. Plan 20 minutes for recording in notebooks and to wrap up with the whole class a day later, after the projects have dried. In addition, plan 20 minutes for the reading.

2. **Preview Part 4**

   Students make plywood from thin strips of wood and glue. Students compare the breakable strength of a craft stick to that of their homemade plywood. The focus question is **How is plywood made?**

3. **Fold newspaper**

   Use one full sheet of newspaper, folded twice, as a work mat for each student.

4. **Prepare for the introduction**

   To introduce the investigation, you will need one plywood sample (from the samples used throughout the module), the *Plywood Production* poster, two or three craft sticks, three thin wood pieces, glue, and a sheet of newspaper. You will also need two or three pieces of 1/4" plywood found in the kit. This part is the only time that the 1/4" plywood is used. These pieces look like the sandwich wood that students will make.

5. **Set up the center**

   Put a folded sheet of newspaper at each student's place. Put 1/4" plywood pieces around the table so students will be able to compare the sandwich wood they make to the real thing. Pile the thin wood pieces in the center of the table where students can easily reach them. Have the white glue ready for distribution.

6. **Plan to read** *Science Resources*: **"Are You a Scientist?"**

   Plan to read "Are You a Scientist?" during a reading period after completing the active investigation for this part.

7. **Plan assessment**

   Here are specific objectives to observe in this part.

   • *Wood pieces can be glued together to make stronger wood.*

   • *Students compare properties of wood.*

FOCUS QUESTION

How is plywood made?

New Word
Say it → See it → Hear it → Write it

Materials for Step 3
- *Thin wood pieces*
- *Newspaper mat*
- *Glue*
- *Paper towels*

GUIDING *the Investigation*
Part 4: *Making Sandwich Wood*

1. **Observe the plywood samples**
   Hold up a plywood sample (the one used throughout the module). Ask if there are any **plywood** trees. Discuss how students think this kind of wood is made. Confirm that plywood is thin **layers** of wood glued together. Tell students that some people like to call plywood "sandwich wood." Ask,

   ➤ *Why do you think people call it sandwich wood?* [Because it is layered, like a sandwich.]

   Show students a craft stick and compare it to one layer of plywood. Ask two or three students to see if they can **break** a craft stick. Then show them the 1/4" plywood and ask them to try to break it.

   Review the poster that shows how plywood is made. Tell students that gluing layers together, or **laminating** layers of wood, makes wood very **strong**. Explain that today, students will be making a piece of 1/4" plywood. Point out that plywood is **laminated**.

2. **Focus question: How is plywood made?**
   Write the focus question on the chart, and read it together.

   ➤ *How is plywood made?*

   Tell students that today they will use glue to stick wood layers together to make a piece of plywood.

3. **Demonstrate the procedure**
   Squeeze a pool of glue about the size of a quarter onto a piece of newspaper and explain to students the procedure for making plywood.

   a. *Get three thin wood pieces.*

   b. *Dip your finger in the glue. Cover one surface of two pieces of wood evenly.*

   c. *Press the two glued surfaces together. Hold them for a count of 20.*

   d. *Add the other piece of wood by following the same procedure.*

4. Make sandwich wood

Send six to ten students to the center. Have each student choose three thin wood pieces. Go around the group and squeeze a quarter-sized pool of glue onto the *far side* of each student's newspaper work mat. Let students begin gluing the pieces together. Remind them to put glue on the entire surface of two thin pieces of wood, then hold the pieces together for a count of 20, and repeat this procedure for the third piece of wood.

5. Add vocabulary to the word wall

As students offer their observations, add any new or important vocabulary to the word wall.

6. Prepare for the next group

Find a location where the wood can dry overnight. Label students' pieces of plywood by writing their names on the wood.

Replace the newspaper for the next group of students, or use the same sheets turned inside out, if they are not covered with too much glue.

7. Review vocabulary

Review key vocabulary added to the word wall. Here's a suggested cloze review. Students answer chorally.

➤ *Thin layers of wood glued together are called _____ .*

S: Plywood (or sandwich wood).

➤ *Another word for gluing layers is _____ .*

S: Laminating.

➤ *Craft sticks do this when you bend them, but plywood does not _____ .*

S: Break.

➤ *Laminated plywood is _____ .*

S: Strong.

Materials for Step 4
- *Newspaper mats*
- *Glue*
- *Thin wood pieces*
- *Paper towels*
- *1/4" plywood pieces*

break
laminated
laminating
layer
plywood
strong

B R E A K P O I N T

FOCUS CHART

How is plywood made?

We glued _____ to make plywood.

We glued layers of wood together to make plywood.

8. **Answer the focus question**

After students' plywood pieces have dried overnight, have students retrieve their pieces and compare them to the plywood samples in the kit. Guide a discussion by asking if plywood grows on plywood trees. Restate the focus question.

➤ *How is plywood made?*

Tell students you have a strip of paper with the question written on it. Review how to glue the strip into the notebook.

Ask students to answer the focus question in drawings and/or words.

Here are additional questions you might discuss with students.

➤ *How is sandwich wood different from sawdust wood?*

➤ *Why do you think people make plywood?* [To make strong lumber.]

READING *in Science Resources*

9. Read "Are You a Scientist?"

Students have been using scientific processes in their investigation of wood. This article reinforces students' roles as scientists by describing the many activities of both scientists and children.

Ask students what a scientist is. Continue the discussion by asking what scientists do and how they act. Brainstorm a list of ideas. Introduce the title of the article, and explain that the article will explain what a scientist is and does. Ask them to listen for new ideas to add to the brainstorming list.

Read the article aloud, pausing to discuss key points.

10. Discuss the reading

Discuss the reading, using these questions as a guide.

➤ *What new ideas can we add to our brainstorming list?*

➤ *What is the most important thing that a scientist does?* (Have students explain their choices.)

➤ *When we study wood, how are we acting like scientists?*

Are You a Scientist?

Scientists ask questions.

9

WRAP-UP

11. Share notebook entries

Conclude Part 4 by having students share notebook entries. Ask students to open their science notebooks to the most recent entry. Read the focus question together as a class.

➤ *How is plywood made?*

Ask students to pair up with a partner to

- share their answers to the focus question;
- explain their drawings.

TEACHING NOTE

Refer to the teacher resources on FOSSweb for a list of appropriate trade books that relate to this module.

INTERDISCIPLINARY EXTENSIONS

Art Extensions

- **Make pictures with sawdust, shavings, and twigs**
  Students can add texture to their drawings by spreading a thin layer of glue and sprinkling on sawdust, shavings, or twigs.

- **Draw with charcoal**
  Explain to students that charcoal is wood that has been changed by burning it. Let students experiment with charcoal drawing.

- **Add to the wood center**
  Bring in sawdust from different kinds of wood, a few wood scraps, sandpaper, and woodworking tools to add to the center for students to explore.

Social Studies Extension

- **Visit a woodworker**
  Contact a woodworker, and plan a field trip or invite the person to class. You might check with the local middle school or high school to see if they have a woodworking class. Ask the woodworker to demonstrate the safe use of tools to shape wood. This is an opportunity to introduce students to careers in construction using wood.

TEACHING NOTE

Review the teacher resources on FOSSweb for community- and career-related extensions.

Home/School Connection

Students investigate whether a craft stick will become waterlogged and sink. After a few days, they compare it to a dry craft stick.

Make copies of teacher master 14, *Home/School Connection* for Investigation 2. Send it and two craft sticks home with students after Part 2.

TEACHING NOTE

Families can get more information about Home/School Connections on FOSSweb.

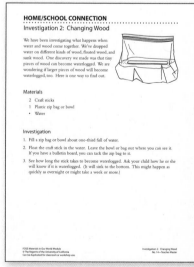

No. 14—Teacher Master

PURPOSE

Content

- Paper has many observable properties.

- People make paper from wood.

- Many objects are made from paper. The properties of different papers determine their uses.

- Some kinds of paper absorb water, while others do not. Some paper changes when soaked in water. Some paper breaks down into small fibers.

- Paper, a resource, can be reused, recycled, and fabricated.

Scientific Practices

- Observe and compare properties of several kinds of paper.

- Determine the usefulness of different kinds of paper for writing and drawing.

- Investigate how paper interacts with water.

- Communicate observations of different kinds of paper orally and through drawings.

	Investigation Summary	Time	Focus Question
PART 1	**Paper Hunt** Students observe and compare the properties of ten kinds of paper. They go on a paper hunt, looking for a sample that matches one that they are given. Students place labels around the classroom to highlight all the items in their environment made of paper.	**Introduction and Hunt** 30 minutes **Notebook** 15 minutes **Reading** 20 minutes	What is made of paper?
PART 2	**Using Paper** Students use crayons, pencils, and marking pens to explore and compare the properties of paper that make it suitable or unsuitable for writing and drawing. Students fold paper and compare the properties of paper that allow it to be folded.	**Introduction** 5 minutes **Center** 20 minutes **Notebook** 20 minutes	What makes paper good for writing? What makes paper easy to fold?
PART 3	**Paper and Water** Students drop water on ten different paper samples and observe and compare the results. They submerge the paper in water and let it dry to see if the paper changes in any way. Students decorate paper flags and hang them on a string outdoors to observe the paper over time.	**Introduction/ Center** 30 minutes **Notebook** 20 minutes **Next Day** 20 minutes	What happens when water gets on paper?
PART 4	**Paper Recycling** Students are introduced to papermaking and recycling. They shake toilet tissue and water in a bottle to make a pulp and then form it into a new sheet of paper. Students discover that the new paper has many of the properties of the original paper and also has some very different properties.	**Introduction** 10 minutes **Center** 20 minutes **Notebook** 20 minutes	How can new paper be made from old paper?
PART 5	**Papier-Mâché** Students use wheat paste (flour and water) to mold strips of newspaper over a small container. They use this papier-mâché technique to change the paper from flexible to stiff and strong so it will keep a shape.	**Introduction** 5 minutes **Center** 30–40 minutes **Next Day/ Notebook** 20 minutes	How can paper be made strong to form a bowl?

At a Glance

Content	Writing/ Reading	Assessment
• Paper has many observable properties. • Many objects are made from paper. • People make paper from wood. Wood is a resource that comes from trees.	**Science Notebook Entry** Draw or write words to answer the focus question. **Science Resources Book** "The Story of a Box"	**Embedded Assessment** Teacher observation
• Paper has many observable properties. • The properties of some kinds of paper make the paper useful for writing or drawing or folding.	**Science Notebook Entry** Draw or write words to answer the focus questions.	**Embedded Assessment** Teacher observation
• Paper has many observable properties. • Water can soak in or form beads on paper. • Paper can be changed by soaking it in water.	**Science Notebook Entry** Draw or write words to answer the focus question.	**Embedded Assessment** Teacher observation
• People make paper from wood. • Wood is a resource that comes from trees. • Old paper can be broken down and made into new paper.	**Science Notebook Entry** Draw or write words to answer the focus question.	**Embedded Assessment** Teacher observation
• Paper, a resource, can be reused, recycled, and fabricated. • Paper can be changed into a material with new properties. • Materials can be mixed together to form a mixture with different properties.	**Science Notebook Entry** Draw or write words to answer the focus question.	**Embedded Assessment** Teacher observation

BACKGROUND *for the Teacher*

Paper is a simple material produced in a straightforward way. All that is needed is a source of cellulose **fibers**, some water, and a screen. Humans have made paper for centuries, continually improving and advancing the technology in little ways.

Cellulose is the wonder fiber at the root of the paper story. Cellulose is a long molecule found in the stems, branches, and leaves of most plants. It is a member of the carbohydrate family, but, unlike sugar and starch, cellulose resists digestion by humans (and most other animals, for that matter). When cellulose-rich material, such as a tree trunk, is smashed, pounded, ripped, churned, and thrashed with water, the fibers separate and are suspended in the water. The cellulose-fiber mush is **pulp**, and the factories that produce it are known as pulp mills.

If the pulp is agitated to keep it uniformly distributed in the water, a **flat** screen can be dipped into the soup and lifted straight up. In the process, a thin layer of cellulose fibers collects on the screen. That's raw paper.

The sheet of paper is finished by **rolling**, pressing, and **drying**. In the process of pressing and drying, the cellulose fibers flatten and bond to one another strongly. No other fiber bonds as effectively without the addition of a binder (glue) of some kind. The same process can be applied to other fibers, such as wool or fur, but the product is felt, not paper. Cellulose is the key to paper.

> *"Paper is one material that is everywhere."*

What Is Made of Paper?

Paper is one material that is everywhere. It is not unreasonable to assume that the average person comes into contact with ten kinds of paper every day, and perhaps as many as 50 kinds in a month. Paper has invaded virtually every aspect of modern life. The traditional information and communication industries rely heavily on printing—books, magazines, newspapers, signs, and labels—and most printing is done on some kind of paper.

Paper was originally developed as a flat, durable surface for writing and drawing—for recording the events, ideas, plans, and visions of humanity. Over the years, however, paper has become much more than a two-dimensional surface for writing. It is an important fabrication material that is prominent in every aspect of life in contemporary societies. Paper is used in housing, food preparation, packaging, shipping, storing, protecting consumer goods, information exchange, homemaking, and just about everything else as well. Paper provides a fast-paced society with convenience and efficiency: paper plates and cups, paper bags to carry purchases, paper wrappings for fast food, **paper towels**, paper tissues, and paper diapers. Paper may not make the world go around, but it does cover it fairly well in a **thin** cellulose wrapping.

There are interesting specialty papers. Often the special twist is an additive to the paper rather than a novel technology for producing the paper itself. **Waxed paper** is totally infused with a dense wax that prevents, or at least delays, the absorption of water. Before the advent of plastic films and sheets, waxed paper was the material of choice for covering and wrapping materials to prevent them from drying out. Plastic- and metal-coated papers are used where moisture and/or light must be excluded from a package.

On the opposite end of the scale are papers that have been developed to absorb liquids quickly and efficiently. Paper towels, toilet tissue, **facial tissue**, and **blotters** have an open construction and are soft, allowing liquids to move into the paper quickly and holding them there effectively.

There is also a host of other specialty papers. Crepe paper stretches, and tracing paper is nearly transparent. Flash paper burns in an instant without any ash, and multiwall **corrugated cardboard** is as sturdy as lumber. The list goes on.

Even though the variety is immense, paper can be grouped into four main categories: **newsprint**, **chipboard**, absorbent paper, and strong paper.

Newsprint is inexpensive to produce and not very durable. The individual fibers in this paper are short, making it easy to **tear**. It is used for products that have a short life expectancy. Newsprint is made from waste products such as sawdust and reclaimed paper. We encounter newsprint most often in the form of newspapers, advertising papers, comic books, and other inexpensive printed products. Paper used in schools for writing and drawing is often newsprint. Newsprint is thin, making it easy to **fold**. It has a smooth texture and is absorbent, making it good for writing and drawing.

Chipboard is also inexpensive to produce and built for function. We see it all the time in the form of packages. The familiar lightweight cardboard box that contains breakfast cereal, cornstarch, crackers, and a multitude of other products is made from this **thick** paper. The color is a clue to its identity: chipboard is made almost exclusively from **recycled** newspaper. The gray-to-tan color indicates that the material was not bleached in the process of being reconstituted. Chipboard is thick, making it hard to fold. But it has texture, and if there isn't a shiny coating on it, it is good for writing as well.

One useful property of paper is that it can be cut into complex shapes quite easily, either by hand or with mass-production machinery. Furthermore, paper can be folded sharply and precisely without damage to the integrity of the material. When the advantages of light weight and low cost are also considered, paper is the material of choice for many projects.

Many paper products are made by cutting, folding, and gluing. Envelopes and bags are familiar products made in this fashion. Cardboard is not always thought of as paper, but it is. The chipboard used for cereal boxes, shoe boxes, and many other containers is thick, crude paper made from recycled newspaper. Such containers provide protection for delicate materials and make many products easier to handle and display. The size, variety, and methods of fabrication for such products seem endless.

Stronger paper containers are made from corrugated cardboard. Corrugations are wrinkles, or folds, and sandwiching a corrugated sheet of paper between two flat sheets of paper produces a very strong, rigid material for fabrication of heavy-duty cartons. Modern systems of commerce are highly dependent on corrugated paper products for moving consumer goods across town, across the country, and around the world.

What Happens When Water Gets on Paper?

Absorbent papers have lots of space between the fibers. Liquids soak into them rapidly, and gases and liquids pass through them easily. Paper towels, coffee filters, and facial tissues are examples of this kind of paper.

Most of the remaining kinds of paper fall into the strong paper category. Strong papers—writing paper, typing paper, paper plates, waxed paper, magazine paper, corrugated cardboard, shopping bags, and **kraft paper**— are smooth and durable. Kraft paper, heavy and often brown, finds extensive application where rough or extended use is expected, such as in shipping cartons, wrapping paper, envelopes, and shopping bags.

When water gets on strong paper, it may take longer to soak in. And in some cases such as waxed paper, the water beads up and doesn't soak in.

How Can New Paper Be Made from Old Paper?

It might be assumed that if paper were put back into water and agitated a bit, it would be reduced again to fibers. This is not usually the case, however; paper holds together. Putting paper into water usually softens it, makes it pliable, and may even cause it to fall into pieces, but not into individual fibers. The cellulose bonds are surprisingly strong.

Even so, paper is an excellent material to recycle once it has outlived its usefulness in its original form. Old paper can be subjected to the same brute forces previously described, and the fibers can once again be isolated, ready to be made into a new sheet of paper. The cellulose fibers usually are shorter as a result of being pounded a second or third time, so the resulting recycled paper may not be as stiff or resist tearing quite as well as new paper, but for most applications, recycled paper is fine.

When paper breaks into little pieces of paper rather than into fibers, it can be reconstituted into new crude paper called chipboard. This is the thin gray cardboard used to make cracker boxes, shoe boxes, the backs of writing tablets, and so on. The main component in chipboard is newspaper.

Say it
Write it
New Word
See it
Hear it

Across
Around
Bend
Blot
Bumpy
Chipboard
Construction paper
Corner
Corrugated cardboard
Crease
Drop
Dry
Facial tissue
Fiber
Flat
Flip
Flour
Fold
Half
Kraft paper
Mold
Newsprint
Over
Paper
Paper towel
Papier-mâché
Pattern
Pulp
Recycling
Rolling
Slick
Stiff
Strip
Submerge
Tagboard
Tear
Thick
Thin
Waxed paper
Wet
Wheat paste

How Can Paper Be Made Strong to Form a Bowl?

Lamination is the process of laying down a number of layers of material with an adhesive to hold the layers together. The process usually produces a product that is stronger than a sample of the base material of equal thickness. For example, a plywood board is stronger than a single board of the same thickness. The same is true of paper. The strip papier-mâché process, often done with newspaper and wheat paste (flour and water), produces a thin, rigid material of surprising strength. Because paper becomes limp and pliable when it is wet with paste, it can be used to form delicate little bowls, costume jewelry, or furniture. Papier-mâché is used to make festive piñatas and heads for hand puppets.

Understanding the properties of paper foldability, joinability, absorbency, and so on is the first step toward being able to use these materials in a variety of ways to make things. When science is put to work for people, it is called technology. In this investigation, students are exposed to a variety of techniques for transforming paper into useful objects.

Paper makes a good box because it is light and can be folded. Paper is good for notebooks because it is a smooth writing surface. Young children know these things because they have had these experiences with paper in their personal lives.

The next level of understanding of paper comes when students themselves make boxes, bowls, and books. Through such enterprise, the worker has to integrate scientific knowledge of properties of materials with technology for transforming and manipulating those materials into products. Students will be budding paper crafters in this investigation, as they learn valuable lessons in transformation.

Folding paper into three-dimensional shapes creates useful products for holding things such as beverages, cookies, and building blocks. It is important for early-childhood students to be introduced to as many kinds of paper containers as possible, and to have opportunities to take them apart and put them back together to see how they are made. Taking things apart and putting them back together is one of the cornerstones of science.

TEACHING CHILDREN *about Paper*

Paper is a much broader category of materials than most early-childhood students realize at first. Cardboard boxes big enough to hide in, a tissue to wipe a drippy nose, paper on an easel for painting, and the box containing a favorite breakfast cereal might be made of completely different materials as far as young children are concerned. However, it is important for students to start finding the similarities in all these materials and to understand that these materials all have similar origins and properties.

Science is a child's area of study. Children are naturally good at it. The subject is familiar—the world—and the approach is discovery. So set the table (in this case with a paper plate), and watch students enjoy the feast.

The **conceptual flow** for Investigation 3 starts when students investigate ten samples of a new material, **paper**, in Part 1. They describe the **properties** of the different kinds and learn their names and how to identify them based on their **softness**, **flexibility**, **thickness**, and **color**.

In Part 2, students test paper to learn how each kind of paper can be **used**, finding which papers are good drawing and **writing surfaces**, and which can be **folded** effectively.

In Part 3, **paper and water interact**. Students put drops on paper surfaces and find that water **soaks in** or **beads up.** When students plunge paper samples into water, they find that paper **sinks**, and some kinds **disintegrate**.

In Parts 4 and 5, students explore **technologies** to **recycle** and **reuse** paper. They shake toilet tissue in water to break it into fibers and craft it into a new piece of paper (**papermaking**). They tear paper into strips and use adhesive to craft the strips into a bowl, using **papier–mâché**. In extension activities, they **fabricate** drinking **cups**, **hats**, **envelopes**, and woven **mats** from paper strips and patterns.

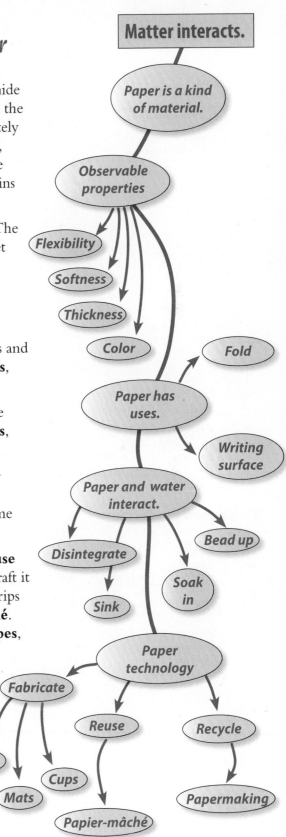

No. 3—Teacher Master

PAPER LABELS

No. 15—Teacher Master

MATERIALS *for*
Part 1: *Paper Hunt*

For each student

1–2 Paper labels (See Step 6 of Getting Ready.)
1 *FOSS Science Resources: Materials in Our World*
  - "The Story of a Box"

For the class

10 Fluted containers
12 Index cards
32 Samples of each kind of paper
  - Chipboard (cereal-box material)
  - Corrugated cardboard
  - Corrugated paper
  - Kraft paper
  - Newsprint
  - Tagboard
  - Waxed paper (See Step 3 of Getting Ready.)
  - Paper towel ★
  - Construction paper, white ★
  - Facial tissue, white ★
  - Removable tape
2 Zip bags, 1 L
❏ 1 Teacher master 3, *Focus Questions B*
❏ 1 Teacher master 15, *Paper Labels*
1 Big book, *FOSS Science Resources: Materials in Our World*

For assessment

❏ • *Assessment Checklist*

★ Supplied by the teacher.　　❏ Use the duplication master to make copies.

GETTING READY *for*

Part 1: *Paper Hunt*

1. Schedule the investigation

Part 1 is a whole-class activity. Plan 30 minutes for the introduction and paper hunt, 15 minutes for recording in notebooks, and 20 minutes for the reading.

2. Preview Part 1

Students observe and compare the properties of ten kinds of paper. They go on a paper hunt, looking for a sample that matches one that they are given. Students place labels around the classroom to highlight all the items in their environment made of paper. The focus question is **What is made of paper?**

3. Prepare paper samples

Six of the ten paper samples are in the kit, already cut into 10-centimeter (cm) squares. These include

- Chipboard
- Kraft paper
- Corrugated cardboard
- Newsprint
- Corrugated paper
- Tagboard

In the kit, you will find a roll of waxed paper. You will need to get white or neutral colors of the following three papers.

- Construction paper
- Facial tissue
- Paper towels

Keep the uncut rolls of waxed paper and the sheets to show students in Step 6 of Guiding the Investigation. Use scissors or a paper cutter to cut fifty 10-centimeter (cm) (4") squares of each of these last four kinds of paper. You will need about 20–22 squares of each kind of paper for this part, depending on class size. Keep all the samples together for Parts 2–3.

4. Prepare for paper distribution

Each kind of paper will be placed in its own fluted container. Place at least 16 squares of paper in each of the ten containers. Set the ten containers on a table where students can file by and pick up one of each kind of paper.

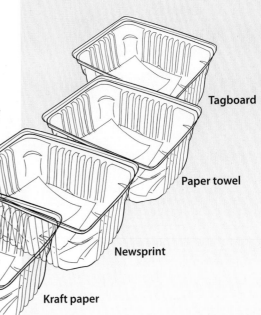

Tagboard

Paper towel

Newsprint

Kraft paper

5. Prepare for the paper hunt

Assemble two identical sets of matched paper samples—one set to place around the room and one set to distribute to students. If you have 20 students in your class, each set is made by putting two samples of each of the ten kinds of paper in a zip bag (20 samples total in each set). Modify the number of samples in each set based on the number of students present for the paper hunt.

While students are out at recess or before they come to school, place *one set* of paper samples around the classroom so that students will be able to find them later. The samples should be in plain sight. Place enough samples around the room for each student to find one match during the hunt.

Have the other set of samples ready to distribute to students at the time of the hunt.

6. Prepare for labeling

Each student will need one or two labels that read, "This is made of paper." Photocopy teacher master 15, *Paper Labels*, and cut the labels apart.

Tear off small pieces of *removable* tape and stick them to the side of a table so they protrude over the edge. Students will use these pieces of tape to attach their labels to things in the classroom that are made of paper. (Removable tape can be easily removed from posters and bulletin-board displays without damage to paper products.)

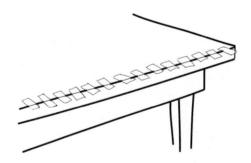

7. Prepare paper sorting mats

At the end of this part, students will put their paper samples in the appropriate containers. To reinforce the names of the papers and the words that describe their properties, prepare sorting mats to place by the containers for this final sorting.

tissue	kraft
white	brown
thin	stiff

8. **Plan to read *Science Resources*: "The Story of a Box"**

Plan to read "The Story of a Box" during a reading period after completing the active investigation for this part.

9. **Plan assessment**

There are four objectives that can be assessed at any time during any of the parts of this investigation.

What to Look For

- *Students ask questions.*

- *Students conduct simple investigation.*

- *Students communicate observations orally.*

- *Students incorporate new vocabulary.*

Here are specific objectives to observe in this part.

- *Paper has many observable properties.*

- *Many objects are made from paper.*

- *Wood is a resource that comes from trees; people make paper from wood.*

- *Students compare properties of paper.*

Focus on a few students each session. Record the date and a + or − on the *Assessment Checklist*.

FOCUS QUESTION

What is made of paper?

GUIDING *the Investigation*
Part 1: *Paper Hunt*

1. **Introduce paper**
   Call students to the rug. Hold up a few pieces of paper from the containers and ask students if they can tell what the pieces are. Confirm that the containers hold different kinds of **paper**.

2. **Form groups and explore the paper samples**
   Arrange students into four or five groups on the rug. Assign two students from each group to file past the containers of papers to pick up one sample from each. Have those students place the two sets of samples on the rug in the center of each group for all students to observe. Visit each group to make sure everyone has a chance to see and touch all the samples.

3. **Name the paper samples**
   Ask students to choose two samples and hold one in each hand.

   Hold up a sample of **kraft paper**. Describe some of its properties, and identify it by name. Ask students to hold their paper samples in the air if they are holding kraft paper. Continue to name and match with the nine remaining kinds of paper. Write the paper names on the word wall. When you've finished the identification activity, collect the samples.

Materials for Step 4
- *Sets of 10 paper samples (chipboard, corrugated cardboard, corrugated paper, kraft paper, newsprint, tagboard, waxed paper, construction paper, facial tissue, paper towel)*

4. **Explain the paper hunt**
   Show students the set of paper samples for the hunt. Explain that each student will get one piece of paper. Tell students that their job is to find one (and only one) piece of paper in the room that is the same as their sample. When they find the one that matches, they should bring the two pieces of paper back to the rug and sit down.

5. **Begin the paper hunt**
   Give each student a sample of paper. Remind students to look at and feel the paper carefully to make sure that they bring back two samples that are exactly the same—some of the papers look very similar.

   As students return to the rug, check that their samples match. If they don't, help the student discover the difference and then let him or her search for an exact match.

TEACHING NOTE

If students can't find a match and seem frustrated, direct them to the containers of paper to find a match.

6. **Discuss uses and packages**
   Gather students at the rug. Show them the rolls of paper you used to prepare some of the samples (waxed paper, paper towels, and facial tissue). Discuss each of the different kinds of paper and what it is used for. Call out the name of each kind of paper, and have students return their samples to the proper containers.

Materials for Step 6
- *Roll of waxed paper*
- *Roll of paper towels*
- *Facial tissue*

7. Focus question: What is made of paper?

Write the focus question on the chart, and read it together.

➤ *What is made of paper?*

8. Label the classroom

Hold up a "This is made of paper." label, and read it aloud. Tell students that everyone will get a label to tape onto something in the room that is made of paper.

Demonstrate how to take a piece of removable tape from the edge of the table, place it on the label, and put it on a paper object. Give each student one or two labels, and let students begin.

9. Review vocabulary

When students have finished labeling objects in the room, call them to the rug and review paper names. Here's a suggested cloze review. Students answer chorally.

➤ *Most cereal boxes are made from _____ .*

S: **Chipboard.**

➤ *A wavy layer of paper is called _____ .*

S: **Corrugated paper.**

➤ *Most large boxes are made from _____ .*

S: **Corrugated cardboard.**

➤ *Paper that has a waxy finish is called _____ .*

S: **Waxed paper.**

➤ *Newspapers are printed on _____ .*

S: **Newsprint.**

➤ *The paper we use for cleaning up spills is a _____ .*

S: **Paper towel.**

➤ *The colored paper we use for art projects is _____ .*

S: **Construction paper.**

➤ *This kind of paper is used for wiping noses.*

S: **Facial tissue.**

➤ *This kind of paper is used for paper bags.*

S: Kraft paper.

➤ *This kind of paper is used for making posters.*

S: **Tagboard.**

Materials for Step 8
- *Paper labels*
- *Removable tape*

This is made of **paper.**

Say it → See it → Hear it → Write it → **New Word**

chipboard
construction paper
corrugated cardboard
corrugated paper
facial tissue
kraft paper
newsprint
paper
paper towel
tagboard
waxed paper

E L   N O T E

Hold up the sample of each paper as you say the name. You can also tape the paper samples to the word wall.

FOCUS CHART

What is made of paper?

_____ *is made of paper.*

10. Answer the focus question

Restate the focus question.

➤ *What is made of paper?*

Tell students you have a strip of paper with the question written on it. Review how to glue the strip into the notebook.

Ask students to answer the focus question in drawings and/or words.

WRAP-UP/WARM-UP

11. Share notebook entries

Conclude Part 1 or start Part 2 by having students share notebook entries. Ask students to open their science notebooks to the most recent entry. Read the focus question together as a class.

➤ *What is made of paper?*

Ask students to pair up with a partner to

- share their answers to the focus question;
- explain their drawings.

READING *in Science Resources*

12. Read "The Story of a Box"

This article extends students' learning by discussing the process used to create a common paper object—a cardboard box.

Introduce the title, "The Story of a Box." Ask students if the title reminds them of another article they have read. If students do not recall "The Story of a Chair," turn to page 3 and ask students if they remember the story of the chair. Ask what the article told them. Ask what they think "The Story of a Box" will tell them. Ask students to predict how a box is made.

Read the article aloud. Pause to discuss key points in the article, to review the pictures, and to make predictions.

13. Discuss the reading

Discuss the reading, using these questions as a guide.

➤ *What is a box made from? Where does it begin?*

➤ *How does a tree become cardboard?*

➤ *Why is cardboard a good material to use to make a box?*

14. Extend the reading

Write the following cloze paragraph on chart paper:

The Story of a Box

First, a _____ is chopped _____.

Next, the _____ is made into _____.

Then, the _____ is made into _____.

Next, the _____ is made into _____.

Then, the _____ is made into _____.

Finally, the _____ is made into a _____.

Write each of the following 12 words on a separate index card: *tree, down, tree, sawdust, sawdust, pulp, pulp, paper, paper, cardboard, cardboard, box.* Post the chart paper on the board, and pass out the index cards to different students. As a class, read the paragraph and fill in each blank with the appropriate word. Have students post their index cards on the chart.

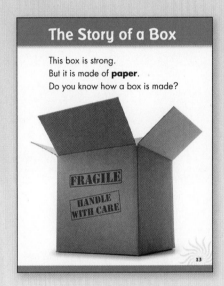

The Story of a Box

This box is strong.
But it is made of **paper**.
Do you know how a box is made?

FRAGILE

HANDLE WITH CARE

13

MATERIALS *for*

Part 2: *Using Paper*

For the class

6 Fluted containers (with paper samples)

- Chipboard samples
- Corrugated paper samples
- Newsprint samples
- Paper towel samples ★
- Tagboard samples
- Waxed paper samples

- Pencils ★
- Crayons and marking pens ★
1 Sheet of standard white paper, 22 × 28 cm (8.5" × 11") ★
❏ 1 Teacher master 16, *Center Instructions—Using Paper A*
❏ 1 Teacher master 17, *Center Instructions—Using Paper B*

For assessment

- *Assessment Checklist*

★ Supplied by the teacher.   ❏ Use the duplication master to make copies.

No. 16—Teacher Master   *No. 17—Teacher Master*

GETTING READY *for*

Part 2: *Using Paper*

1. Schedule the investigation

This part requires 20 minutes at the center for each group of six to ten students. Plan 5 minutes to introduce, 15 minutes for recording in notebooks, and 5 minutes to wrap up.

2. Preview Part 2

Students use crayons, pencils, and marking pens to explore and compare the properties of paper that make it suitable or unsuitable for writing and drawing. Students fold paper and compare the properties of paper that allow it to be folded. The focus questions are **What makes paper good for writing? What makes paper easy to fold?**

3. Plan for the introduction

Gather a set of six paper samples to share with the class (one of each kind listed in the Materials section). Have a crayon, marking pen, and pencil on hand.

4. Review paper samples

Look over the paper samples to make sure you have enough for each student to have one sample of each kind of paper. For waxed paper, you should have two samples for each student, one for writing and one for folding.

5. Set up the center

Place the containers of paper samples in a location convenient to the center so that students can pick up paper samples as needed. Have pencils, crayons, and marking pens ready. Use a standard 22 × 28 cm (8.5" × 11") sheet of white paper to introduce the concept of folding paper in half.

6. Plan to roll up sleeves

Ink on waxed paper beads up and can cause a mess. Help students roll up their sleeves before they use the marking pens. Remove the waxed paper after students have tried to mark it. Provide new samples of waxed paper for the folding activity.

7. Plan assessment

Here are specific objectives to observe in this part.

- *Paper has observable properties.*

- *Compare properties of paper.*

Focus on a few students each session. Record the date and a + or − on the *Assessment Checklist*.

FOCUS QUESTION

What makes paper good for writing?
What makes paper easy to fold?

Materials for Steps 1–2
- *Containers of paper samples*
- *Pencils, crayons, and marking pens*

TEACHING NOTE

It is important to structure Steps 3 and 4 carefully to model observing and comparing techniques that students should use when they work on their own to investigate the other papers.

GUIDING *the Investigation*
Part 2: *Using Paper*

1. **Introduce the investigation**

   Call students to the rug. Show them the six kinds of paper they will use in this activity. Hold up a pencil, crayon, and marking pen. Tell students,

   Today your challenge is to find out which kinds of paper are good for writing on.

   Send six to ten students to the center.

2. **Compare tagboard and paper-towel samples**

   When students are settled at the center, place the container of tagboard samples and the container of paper-towel samples on the table. Ask students to take one of each. Discuss how the samples are alike and how they are different.

3. **Test paper samples with pencils and crayons**

   Tell students to use a pencil to make some marks on each paper sample. Discuss which paper provided the best surface for writing with a pencil. Follow the same procedure, using a crayon.

   Let students know that it is OK if the paper rips or gets a hole in it. That's valuable information to have when making a decision about what kind of paper to use for a specific purpose.

4. **Test paper samples with marking pens**

   Tell students to make marks on both paper samples with a marking pen. Compare the results of the marks on the two kinds of paper. Discuss how the ink from the marking pen soaked into the paper towel. [The paper towel absorbed the ink.] Compare this to the tagboard, which did not absorb much ink.

5. **Investigate other paper samples**

   Show students the four other kinds of paper. Let them mark these papers, using the pencils, crayons, and marking pens. Monitor the use of waxed paper and marking pens. Discuss results with students as they explore, comparing the different papers. Ask,

   ➤ *Is this a good paper for writing a note? Why or why not?*

   ➤ *Does the ink from the marking pen soak into this paper or stay on the top? How do you know?*

   ➤ *What makes paper easy to write or draw on? What makes it hard to write or draw on?*

➤ *Find two papers that feel the same when you write on them. How are they the same?*

➤ *Find two papers that feel different when you write on them. How are they different?*

➤ *How did you feel when you were writing on the corrugated (**bumpy**) paper?*

6. **Focus question: What makes paper good for writing?**
Write the first focus question on the chart, and read it together.

➤ *What makes paper good for writing?*

7. **Discuss findings about paper and writing**
Ask students to hold up paper samples in response to these questions.

➤ *Which paper is the easiest to write on? Why do you think so?*

➤ *Which paper is the best for writing a note? Why do you think so?*

➤ *Which paper is the most fun to write on? Why do you think so?*

Students usually enjoy writing on the corrugated paper because it's bumpy, so it's fun. They will find the waxed paper difficult to write on; usually the marking-pen ink will bead up and not stay on the paper.

8. **Add vocabulary to the word wall**
As students offer their explanations and observations, add any new or important vocabulary to the word wall. Let students be the guides—acknowledge the words they use, and offer new vocabulary as needed, such as, **bumpy**, **slick**, and **tear**.

New Word
Say it
See it
Hear it
Write it

> **E L   N O T E**
>
> **Be sure to review the different possible pronunciations and meanings of tear.**

B R E A K P O I N T

9. **Demonstrate folding paper in half**
Ask students to put all writing tools down while you show them something new. Hold up a sheet of standard white paper. Show students how to **fold** the paper in half, step-by-step. Fold it in half again and again, until you can no longer fold it. Crease each fold firmly before going on to the next fold.

Materials for Step 9
• *Sheet of white paper*

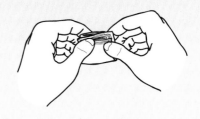

TEACHING NOTE

Folding in half may be a new concept for early-childhood students. The first paper-folding experience in this part is quite structured in order to help students learn this skill. If they already know how to fold, let them pick up all the paper samples at once and begin folding right away.

Say it
Write it **New Word** See it
Hear it

10. Count number of folds

Have students help you count the number of times you were able to fold the paper. Count one fold for each time you open up the paper until you return to the full sheet. Ask if the paper looks the same now as it did before it was folded.

11. Fold newsprint

Have students fold the marked-up piece of newsprint in half once. Make sure they understand the concept of folding in half. Have them continue folding in half until the limit is reached. Have students count the number of folds by unfolding their samples and counting each time they open up a fold.

12. Focus question: What makes paper easy to fold?

Write the second focus question on the chart, and read it together.

➤ *What makes paper easy to fold?*

13. Fold other papers

Ask students to try folding the other paper samples (using the ones they have written on). Continue until students have folded a number of different kinds of paper.

14. Discuss the folding experience

As students fold, guide their discussion by asking questions.

➤ *What kinds of paper are easy to fold, and what kinds are difficult to fold?*

➤ *What makes some kinds of paper easy to fold and other kinds difficult to fold?*

➤ *What does a sheet of paper look like after it has been folded and then opened up?*

➤ *How many times do you think you could fold a facial tissue?*

➤ *Why would people want to fold paper?*

➤ *What could you make by folding paper?*

As students offer their observations, add any new vocabulary to the word wall (**bend, corner, crease, flat, half, thick, thin**).

15. Sort paper samples

When students have finished folding all their samples, ask them to sort the samples into two piles: those that were easy to fold and those that were more difficult to fold. Discuss the properties that affect how easy it is to fold a sheet of paper.

16. Clean up

Each new group will need samples that have not been previously written on or folded. Collect the used samples and save them for future activities, such as making paper collages and paper flags.

17. Review vocabulary

Following the center activities, call students to the rug, and review the key vocabulary. Here's a suggested cloze review.

bend
bumpy
corner
crease
flat
fold
half
slick
tear
thick
thin

➤ *When the ink soaks into the paper, we say the paper _____ the ink.*

S: Absorbs.

➤ *When there are little hills on the paper, we call the paper _____ .*

S: Bumpy.

➤ *Paper that is waxy and doesn't absorb very well is called _____ .*

S: Slick.

➤ *A rip in the paper is called a _____ .*

S: Tear.

➤ *The line made by folding paper is called a _____ .*

S: Crease.

➤ *When you fold the paper down the middle so both sides are the same, it is called folding in _____ .*

S: Half.

➤ *Two edges of a sheet of paper meet at a _____ .*

S: Corner.

➤ *It is hard to fold paper that is _____ .*

S: Thick.

➤ *It is easy to fold paper that is _____ .*

S: Thin.

Discuss these additional words, and write them on the word wall.

- Bend
- Flat
- Fold

EL NOTE

Demonstrate or show examples of each word.

18. Answer the focus questions

Restate the two focus questions.

➤ *What makes paper good for writing?*

➤ *What makes paper easy to fold?*

Tell students you have strips of paper with the questions written on them. Review how to glue the strips into the notebook. Ask students to answer one or both of the focus questions in drawings and/or words.

FOCUS CHART

What makes paper good for writing?

Flat not bumpy. Paper absorbs mark.

What makes paper easy to fold?

Flat and thin.

WRAP-UP/WARM-UP

19. Share notebook entries

Conclude Part 2 or start Part 3 by having students share notebook entries. Ask students to open their science notebooks to the most recent entry. Read the focus questions together as a class.

➤ *What makes paper good for writing?*

➤ *What makes paper easy to fold?*

Ask students to pair up with a partner to

- share their answers to one of the focus questions;
- explain their drawings.

TEACHING NOTE

*See the **Home/School Connection** for Investigation 3 at the end of the Interdisciplinary Extensions section. This is a good time to send it home with students. There are two sheets to send home.*

MATERIALS *for*

Part 3: *Paper and Water*

For each student at the center

1 Dropper

1 Plastic cup

For each group

1 Basin

- Crayons and marking pens ★

For the class

10 Fluted containers (with paper samples)

- Chipboard samples
- Corrugated cardboard samples
- Corrugated paper samples
- Kraft paper samples
- Newsprint samples
- Tagboard samples
- Waxed paper samples
- Construction paper sample, white ★
- Facial tissue samples ★
- Paper towel samples ★
- Clothesline (or string) and clothespins (optional) ★

2 Sponges

- Towel (optional) ★
- String
- Transparent tape
- Water ★
- Newspaper ★

❑ 1 Teacher master 18, *Center Instructions—Paper and Water*

For assessment

- *Assessment Checklist*

★ Supplied by the teacher.   ❑ Use the duplication master to make copies.

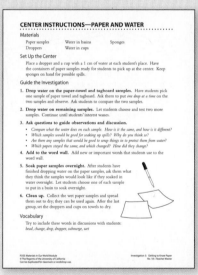

No. 18—Teacher Master

GETTING READY *for*

Part 3: *Paper and Water*

1. **Schedule the investigation**

   This part takes 20 minutes at the center for each group of six to ten students or for the whole class at one time. Plan 10 minutes for the introduction and 20 minutes for recording in notebooks. On the next day, plan a whole-class session of 20 minutes.

2. **Preview Part 3**

   Students drop water on ten different paper samples and observe and compare the results. They submerge the paper in water and let it dry to see if the paper changes in any way. Students decorate paper flags and hang them on a string outdoors to observe the paper over time. The focus question is **What happens when water gets on paper?**

3. **Plan for soaking samples overnight**

   Set the basins, half filled with water, on a table. Each student will put a sample of paper in the basins to soak overnight.

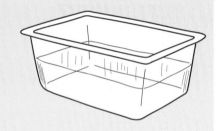

4. **Set up the center**

   Place a dropper and a cup with 1 cm of water at each student's place. Have the ten containers of paper samples ready for students to pick up at the center. Keep a sponge or towel on hand for possible spills.

5. **Plan for drying samples**

   Wet paper samples will need to dry overnight. Set out newspapers on which the samples can be laid, or plan to hang the samples from a clothesline if one is available.

6. **Select your outdoor site**

   Look for a location where you can hang a string of paper flags outdoors: between two trees, two poles, on a fence, or anywhere the flags will be affected by the wind, sunshine, and rain. You might need help from the custodian to get the flags mounted high enough to position them out of mischief's way. Ideally, find a location where students can observe the flags during recess or on their way to and from class. Leave plenty of string at each end for tying the flags securely outdoors.

   See Step 13 of Guiding the Investigation for more details about what you will do outside with students in this part.

7. **Check the site**

   Tour the outdoor site on the morning of an outdoor activity. Do a quick check for unsightly and distracting items.

8. **Prepare a sample string of flags**

   Make a small string of flags ahead of time to show students. Cut a 50 cm piece of string, and tape four or five papers (recycled from Part 2) to the string as flags.

9. **Plan assessment**

   Here are specific objectives to observe in this part.

   • *Water soaks into some kinds of paper. Water forms beads on waxed paper and takes longer to soak in.*

   • *Paper can be changed by soaking it in water.*

   • *The properties of paper determine how it is used.*

   • *Students compare properties of paper.*

   Focus on a few students in each session. Record the date and a + or – on the *Assessment Checklist*.

FOCUS QUESTION

What happens when water gets on paper?

Materials for Step 2
- *Cups of water*
- *Droppers*
- *Containers of paper samples*
- *Sponges*
- *Towel (optional)*

GUIDING *the Investigation*
Part 3: *Paper and Water*

1. **Focus question: What happens when water gets on paper?**
   Call students to the rug. Write the focus question on the chart, and read it together.

   ➤ *What happens when water gets on paper?*

   Ask students what they think might happen when they put drops of water on paper. Listen to several responses, and then tell students they will test to find out.

2. **Drop water on paper samples**
   Send six to ten students to the center with this procedure.

   a. *Start with the paper-towel and tagboard samples.*

   b. *Put only one drop at a time on the samples. Observe carefully.*

   c. *Add more drops, one at a time, and observe.*

   d. *Choose two more paper samples, and repeat the testing.*

3. **Ask questions to guide observations**
   As students drop water on the paper, guide their observations.

   ➤ *Compare what the water does on each sample. How is it the same, and how is it different?*

   ➤ *Which samples would be good for soaking up spills? Why do you think so?*

   ➤ *Are there any samples that would be good to wrap things in to protect them from water?*

   ➤ *Which papers stayed the same, and which changed? How did they change?*

   As students describe their observations, make a note of the key words on the word wall.

4. **Soak paper samples overnight**
   After students have finished dropping water on the paper samples, ask,

   ➤ *What do you think these papers would look like if they were soaked in water overnight?*

Show students the basins of water. Have students choose just one sample of any kind of paper to soak overnight and put it in one of the basins. When all students have put in one sample, you should have at least one piece of each kind of paper soaking in the water.

5. **Clean up**

Collect the remaining paper samples and spread them out to dry; the dried samples can be used again for this part with another group or another class. Set the droppers and cups out to dry.

6. **Review vocabulary**

After students have completed the center activities, call them to the rug. Review the key vocabulary added to the word wall. Here's a suggested cloze review. Students answer chorally.

➤ *When the water soaks in, the paper gets _____ .*

S: **Wet**.

➤ *When the water doesn't soak in, it forms a _____ on the surface.*

S: **Bead**.

➤ *When we put paper in water to soak it overnight and push it down under the water, we _____ the paper.*

S: **Submerge**.

Discuss these additional words and write them on the word wall.

- Change
- Drop

7. **Answer the focus question**

Restate the focus question.

➤ *What happens when water gets on paper?*

Tell students you have a strip of paper with the question written on it. Review how to glue the strip into the notebook.

Ask students to answer the focus question, using drawings and/or words.

B R E A K P O I N T

---

bead
change
drop
submerge
wet

FOCUS CHART

What happens when water gets on paper?

Water makes beads on waxed paper.

Water soaks in other paper.

Materials for Steps 8–9
- *Basins of wet papers*
- *Newspapers or clothesline*

8. **Observe soaked samples**

   Observe the samples that were soaked overnight in water, and discuss the changes that occurred.

9. **Dry the wet samples**

   Put a set of paper samples out to dry on newspapers or on a clothesline. When the samples are dry, ask students to compare the dried samples to a set of samples that has not been soaked.

B R E A K P O I N T

10. **Explore paper outdoors**

    Call students to the rug and explain the outdoor part of this activity. Tell students,

    We put water on several different kinds of paper and observed. We soaked paper in water and set it out to dry. Now we are going to put some paper outside to see how it changes over time.

Materials for Step 11
- *Sample string of flags*

11. **Introduce the paper flags**

    Tell students that they will each select a piece of paper and draw a little picture, using crayons or marking pens. Each student can select the kind of paper he or she wants to use. This will become a paper flag. Students will attach the flags made of different kinds of paper to a string and hang the string outside. Show students the small string of flags that you made ahead of time.

Materials for Step 12
- *Paper samples*
- *Crayons, marking pens*
- *String*
- *Transparent tape*

12. **Decorate paper flags**

    Each student will spend about 5 minutes decorating both sides of a paper flag. As students finish, attach the flags to the string with transparent tape.

13. **Go outdoors**

    Have students help you carry the string of flags outside so that the string doesn't tangle. Hang the string in an area where students can watch the flags at recess. Flag lines can hang between two supports, such as a clothesline, or they can hang down from a single anchor position, such as a second-story window.

    After you hang the flags, gather students in a circle and ask,

    ➤ *Does anyone see anything outdoors that is made of paper?*

    Go on a brief walk and ask students to look for paper. Ask students to consider why outdoor things are not made of paper. This would be a good time to look at the "This is made of wood." labels that were taped on wood outside in Investigation 1 and to observe the condition of the paper labels. Look at a few examples.

14. Return to class

Tell students that they will watch the paper flags over the next few weeks to see if they change. Return to the classroom.

WRAP-UP/WARM-UP

15. Share notebook entries

Conclude Part 3 or start Part 4 by having students share notebook entries. Ask students to open their science notebooks to the most recent entry. Read the focus question together as a class.

➤ *What happens when water gets on paper?*

Ask students to pair up with a partner to

- share their answers to the focus question;
- explain their drawings.

MATERIALS *for*

Part 4: *Paper Recycling*

For each student at the center

- 1 Bottle, clear plastic, with cap, 120 mL
- 1 Sponge
- 1 Self-stick note
- 2 Sheets of newspaper ★

For the class

- 1 Sample of homemade paper (See Step 7 of Getting Ready.)
- 2 Containers, 1/2 L
- 2 Basins
- 10 Screens
- • Waxed paper
- 1 Roll of single-ply toilet tissue ★
- 1 Pitcher or empty 2 L soft-drink bottle ★
- • Water ★
- ❑ 1 Teacher master 19, *Center Instructions—Paper Recycling*

For assessment

- • *Assessment Checklist*

★ Supplied by the teacher.   ❑ Use the duplication master to make copies.

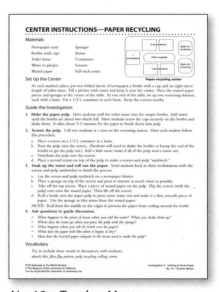

No. 19—Teacher Master

GETTING READY *for*
Part 4: *Paper Recycling*

1. Schedule the investigation
This part requires 20 minutes at the center for each group of five students. Plan 10 minutes to introduce and 20 minutes for recording in notebooks and to wrap up with the whole class.

2. Preview Part 4

Students are introduced to papermaking and recycling. They shake toilet tissue and water in a bottle to make a pulp and then form it into a new sheet of paper. Students discover that the new paper has many of the properties of the original paper and also has some very different properties. The focus question is **How can new paper be made from old paper?**

3. Acquire toilet tissue
Get an inexpensive roll of single-ply toilet tissue. Each student will need about eight squares.

4. Cut the waxed paper
Tear off sheets of waxed paper 15 cm (6") long. Cut the sheets in half so each student gets a piece about 15 × 15 cm (6" × 6").

5. Prepare the newspaper
Each student will need two full sheets of newspaper, folded in half twice, to use as a blotter. (The newspaper can be reused if you allow it to dry overnight.)

6. Set up the paper-recycling center
Work at a table large enough to accommodate five workstations plus two screening stations.

At each student's place, put the two folded sheets of newspaper, a bottle with a cap, and an eight-square length of toilet tissue. Fill a pitcher with water, and keep it near the center. Place the waxed-paper pieces and sponges in the center of the table where students will be able to reach the items.

At one end of the table, set up two screening stations. You will need the two basins, two 1/2 L containers, and ten screens.

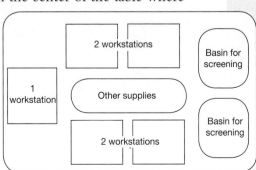

Paper-recycling center

7. **Practice making paper**

   Practice recycling paper, following the technique outlined in Guiding the Investigation, before trying it with students. Save a sample to show students when you introduce the activity.

8. **Plan assessment**

   Here are specific objectives to observe in this part.

   - *Wood is a resource that comes from trees. Paper is made from wood.*

   - *Old paper can be broken down and made into new paper.*

   - *Compare properties of paper.*

   Focus on a few students each session. Record the date and a + or − on the *Assessment Checklist*.

GUIDING *the Investigation*
Part 4: *Paper Recycling*

1. **Introduce paper recycling**
   Call the class to the rug. Ask students to recall where paper comes from. [Wood.] Ask where wood comes from. [Trees.] Have students imagine the trees that were cut to make the paper in their classroom. Tell them,

   *When you use old paper to make new paper, we call that **recycling**. We can save trees by recycling old paper.*

2. **Focus question: How can new paper be made from old paper?**
   Write the focus question on the chart, and have students read it together.

   ➤ *How can new paper be made from old paper?*

   Tell students that they are going to make a small piece of paper. Show them the piece of paper you made (if you have one) and the toilet tissue you used to make it.

3. **Make the paper pulp**
   Send groups of five students at a time to the center. Have students stuff the eight-square lengths of toilet tissue into the empty bottles. Add water to the bottles until they are about two-thirds full. Have students screw the caps securely on the bottles and shake them.

   Tell students to shake their bottles for 3–5 minutes. Have students tell you when the paper has broken down into very small pieces, or **pulp**.

Materials for Step 2
- *Homemade paper*
- *Toilet tissue*

Materials for Steps 3–5
- *Bottles with caps*
- *Screens*
- *Sponges*
- *Newspaper mats*
- *Containers*
- *Basins*
- *Waxed paper pieces*
- *Toilet tissue sheets*
- *Pitcher with water*

4. Screen the pulp

Call two students at a time to the screening stations at the end of the table. Have each student follow this procedure.

a. *Place a screen on a 1/2 L container in a basin.*

b. *Pour the pulp onto the screen. (Students will need to shake the bottles or use the palms of their hands to gently bump the end of the bottle to get all the pulp out.) If all the pulp does not come out, add a little more water.*

c. *Distribute the pulp over the screen.*

d. *Place a second screen on top of the pulp and press down gently.*

5. Soak up the water and roll out the paper

Send students back to their workstations with their pulp-and-screen "sandwiches" to finish. Tell students how to remove the paper from the screens.

a. *Lay the screen-and-pulp sandwich on the newspaper blotter.*

b. *Place a sponge on top of the screen, and press on the screen to remove as much water as possible.*

c. *Carefully lift the top screen off the pressed pulp. Place a piece of waxed paper on the pulp. Flip the screen and pulp over onto the piece of waxed paper, and lift off the remaining screen.*

d. Roll the bottle over the paper pulp like a rolling pin in order to press more water out and make it a thin, smooth piece of paper. Roll from the middle to the edges to prevent the paper from curling around the bottle. Use the sponge to **blot** the water that gets squeezed onto the waxed paper.

6. **Ask questions to guide discussion**
   As students work, ask questions to guide their progress.

   ➤ *What happens to the pieces of tissue when you add the water? When you shake them up?*

   ➤ *Where does the water go when you press the pulp with the sponge?*

   ➤ *What happens when you roll the bottle over the paper?*

   ➤ *What does the paper look like when it begins to dry?*

   ➤ *How does the recycled paper compare to the tissue used to make the pulp?*

7. **Add vocabulary to the word wall**
   As students offer their observations, add any new or important vocabulary to the word wall. Let students be the guides—acknowledge the words they use, and offer new vocabulary as needed.

8. **Dry the paper overnight**
   Label the newly made pieces of paper with self-stick notes. Let the paper dry overnight on waxed paper. Students will glue the paper into their notebooks during the Wrap-Up in Step 11.

Materials for Step 8
- *Self-stick notes*

B R E A K P O I N T

blot
fiber
flip
pattern
pulp
recycling
rolling

9. Review vocabulary

When all students have engaged in the center activities, call them to the rug and review the key vocabulary added to the word wall. Here's a suggested cloze review. Students answer chorally.

➤ *The paper fiber and water mush is called _____ .*

S: Pulp.

➤ *We poured the paper pulp onto a _____ .*

S: Screen.

➤ *We used the bottle like a rolling pin to press the paper pulp. This is called _____ .*

S: **Rolling**.

➤ *We made new paper out of old paper. This is called _____ .*

S: Recycling.

Discuss these additional words and write them on the word wall.

- Blot
- Fiber
- Flip
- Pattern

10. Answer the focus question

Restate the focus question.

➤ *How can new paper be made from old paper?*

Tell students you have a strip of paper with the question written on it. Review how to glue the strip into the notebook.

Ask students to answer the focus question in drawings and/or words.

FOCUS CHART

How can new paper be made from old paper?

Make paper pulp. Pour pulp on a screen. Roll pulp flat. Dry it. This is recycling paper.

WRAP-UP/WARM-UP

11. Share notebook entries

Conclude Part 4 on the next day by having students observe the dried paper and share notebook entries. Ask students to open their science notebooks to the most recent entry. Read the focus question together as a class.

➤ *How can new paper be made from old paper?*

Ask students to pair up with a partner to

- share their answers to the focus question;
- explain their drawings;
- glue their paper samples into their notebooks.

MATERIALS *for*

Part 5: *Papier-Mâché*

For each pair of students at the center

- 2 Containers, 1/2 L
- 2 Plastic cups
- 2 Self-stick notes
- 1 Sheet of newspaper ★

For the class

- • Newsprint (plain) or brown paper towels (from school) ★
- • Flour, 3/4–1 L (3–4 cups) ★
- • Water ★
- • Newspaper ★
- 1 Container, 1/2 L
- 1 Large spoon (with long handle) ★
- 1 Screwdriver or mat knife ★
- 1 Scissors ★
- ❑ 1 Teacher master 20, *Center Instructions—Papier-Mâché*

For assessment

- • *Assessment Checklist*

★ Supplied by the teacher.

❑ Use the duplication master to make copies.

CENTER INSTRUCTIONS—PAPIER-MÂCHÉ

Materials

Newspaper, folded, 1/2 sheets; Newspaper to cover tables; Containers, 1/2 L; Wheat paste; Plastic cups; Self-stick notes

Set Up the Center

Place a half sheet of folded newspaper and a 1/2 L container at each student's place. Pour some wheat paste in a cup for each student. Make sure no student is allergic to wheat.

Guide the Investigation

1. **Tear the newspaper.** Have students tear their folded sheets of newspaper into strips 2.5 cm (1") wide, then into 10 cm (4") lengths. Have students tear the entire sheet of newspaper.
2. **Begin to papier-mâché.** Have students follow these directions.
   a. Place the 1/2 L container upside down.
   b. Cover the container with a thin layer of wheat paste, spreading it evenly with your fingers.
   c. Lay a paper strip over the pasted bowl, and cover it with a layer of paste. Be sure to use enough paste to completely soak the paper, then wipe the excess off.
   d. Follow the same procedure using a second strip of paper, overlapping the first strip already on the container.
   e. Cover the entire outside of the container with several layers of paper strips, down to but not over the rim of the container.
3. **Ask questions to guide discussion.**
   - *What does the wheat paste look like?*
   - *How does the wheat paste feel? How does it feel when it dries on your hands?*
   - *What happens to the paper when it is soaked in paste?*
   - *Is the paper easy or hard to tear when it is soaked in paste?*
4. **Let the papier-mâché dry overnight.** Use self-stick notes to label each student's project on the inside of the container. Set the bowls out to dry.
5. **Add to the word wall.** Add new or important words that students use to the word wall.
6. **Clean up.** Clean the work area for the next students.

Vocabulary

Try to include these words in discussions with students:
across, around, dry, flour, layer, mold, over, papier-mâché, stiff, strip, tear, wheat paste

FOSS Materials in Our World Module
© The Regents of the University of California
Can be duplicated for classroom or workshop use.

Investigation 3: Getting to Know Paper
No. 20—Teacher Master

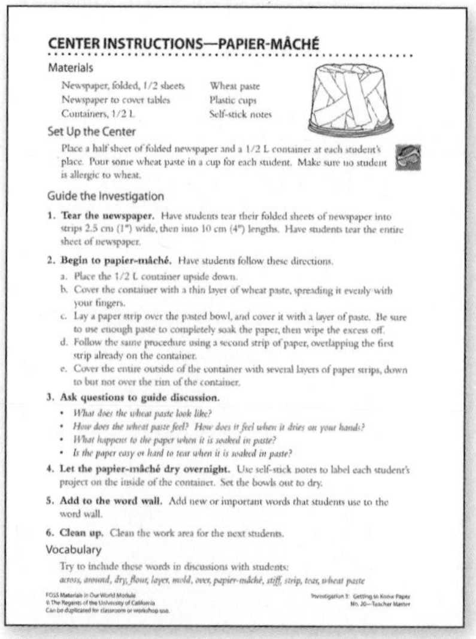

No. 20—Notebook Master

GETTING READY *for*
Part 5: *Papier-Mâché*

1. **Schedule the investigation**

   This part requires 30–40 minutes at the center for each group of six to ten students. Plan a 5-minute introduction and 15–20 minutes the next day after the bowls have dried to record in notebooks and to wrap up the session with the whole class. You will need to plan extra sessions if students make additional layers and decorate their projects.

2. **Preview Part 5**

   Students use wheat paste (flour and water) to mold strips of newspaper over a small container. They use this papier-mâché technique to change the paper from flexible to stiff and strong so it will keep a shape. The focus question is **How can paper be made strong to form a bowl?**

3. **Make wheat paste**

   Mix equal amounts of flour and water to make wheat paste. It should be about the consistency of thin pancake batter. Prepare about 2 cups of wheat paste for each group of students.

   Check to see if any students have a wheat allergy. Non-wheat flour, such as rice flour, can be purchased and used. White glue diluted with water will also make a good paste.

4. **Prepare the newspaper**

   For each pair of students, fold one full-size sheet of newspaper in half twice, then tear it in half. Each student gets one folded half, which he or she will tear into strips.

5. **Make a sample**

   Make a papier-mâché bowl following the procedure in Step 4 of Guiding the Investigation. It will take 2–3 days to dry. Have a 1/2 L container, a cup of wheat paste, and a sheet of newspaper ready to demonstrate.

6. **Set up the papier-mâché center**

   Place a half sheet of folded newspaper and a 1/2 L container at each student's place. Pour a small amount of wheat paste into a cup for each student.

7. **Make cleanup easy**

   When students are finished and all the papier-mâché projects are removed from the table, fold up the dry newspaper and scraps and recycle them. Paper covered with paste can be composted. Soak all the containers in water.

▶ **SAFETY NOTE**

Be aware of students with wheat allergies. Rice flour can be used if necessary.

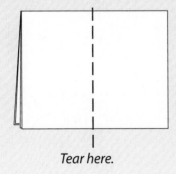

Tear here.

8. **Plan assessment**

Here are specific objectives to observe in this part.

- *Paper can be changed into a material with new properties.*

- *Materials can be mixed together to form a mixture with different properties.*

Focus on a few students each session. Record the date and a + or − on the *Assessment Checklist*.

GUIDING *the Investigation*
Part 5: *Papier-Mâché*

1. Introduce papier-mâché
Call students to the rug. Tell them,

We are going to make a bowl out of paper. We will use a technique called **papier-mâché**. *Papier-mâché is a way to make things out of paper and paste. This is another way to recycle old paper.*

Show students the papier-mâché bowl sample and the 1/2 L container on which it was molded. Show them how the bowl slides on and off the container. Discuss other things that are made from papier-mâché. [Piñatas, puppets.]

2. Focus question: How can paper be made strong to form a bowl?
Write the focus question on the chart, and read it together.

➤ *How can paper be made strong to form a bowl?*

3. Demonstrate tearing the newspaper
Send groups of six to ten students to the center. Show students how to tear thin strips of paper from a half sheet of folded newspaper. The strips should be about 2.5 cm (1") wide. Show students how to *start at the folded edge* and tear all the way down in one fast movement.

Have students tear these long strips into lengths about 10 cm (4") long. Have students tear the long strips before beginning the papier-mâché work. It is too difficult to tear strips when fingers are sticky with wheat paste.

4. Demonstrate the papier-mâché procedure
Demonstrate the procedure.

a. *Place the 1/2 L container upside down.*

b. *Cover the container with a thin layer of wheat paste, spreading it evenly with your fingers. (Explain that wheat paste is a mixture of flour and water.)*

c. *Lay a paper strip on top of the pasted bowl, and cover the strip with a layer of paste. Be sure to use enough paste to completely soak the paper. Wipe off the excess.*

d. *Follow the same procedure, using a second strip of paper, overlapping the first strip already on the container.*

e. *Cover the entire outside of your container with several **layers** of paper strips, down to but not over the rims of the containers.*

FOCUS QUESTION

How can paper be made strong to form a bowl?

Materials for Steps 1–3
- *Papier-mâché bowl sample*
- *Containers, 1/2 L*
- *Half sheets of newspaper*
- *Containers of wheat paste*
- *Self-stick notes*

Start here.

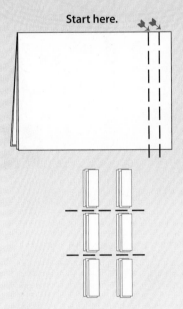

▶ **SAFETY NOTE**

Be aware of students with wheat allergies. Rice flour can be used if necessary. Have students wash their hands with soap and warm water after the activity.

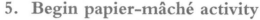

5. **Begin papier-mâché activity**

   As students begin to work, point out the two rims on their containers and tell students not to stick paper over those rims. If the strips are molded over the edges of the rims, the finished product will be difficult to remove when dry.

   Encourage students to cover the whole container with strips in all different directions, including around the container so that the finished product will be sturdy.

   NOTE: You may encounter these problems.

   • Too much paste.

   • Too little paste.

   • Strips too large.

   • Strips laid over container rims. Students should not paste strips over the raised areas of the rims.

   • Reluctance to participate because students think the paste is too messy.

6. **Ask questions to guide discussion**

   As students work, ask,

   ➤ *What does the wheat paste look like?*

   ➤ *How does the wheat paste feel? How does it feel when it dries on your hands?*

   ➤ *What happens to the paper when it gets soaked in the paste?*

   ➤ *Is the paper easy or hard to tear when it is soaked in paste?*

7. **Add vocabulary to the word wall**

   As students offer their observations, add any new or important vocabulary to the word wall.

8. **Dry the papier-mâché overnight**

   Use self-stick notes to label each student's project on the inside of the container. Place the finished papier-mâché bowls on a counter or shelf to dry overnight or longer, depending on the weather. Place each piece upside down so it is resting on the container rims.

 B R E A K P O I N T

9. Lay down additional layers (optional)

After the papier-mâché has dried, have students add another layer of strips to make the bowls stronger. This time use plain newsprint or paper-towel strips and wheat paste to cover the bowls. Use paper towels that don't absorb a lot of water. Brown paper towels from school usually work well. Using plain paper makes it easier for students to see when they have covered the entire bowl, and the print from the newspaper won't show through if students decorate the bowls later.

Materials for Step 9
- *Plain newsprint or brown paper towels*

10. Remove the bowls from the molds

Let the bowls dry for a day or two. When they are completely dry, remove them from the **molds**. This should be a teacher job, but students should watch. Use care with the tools. Slip a screwdriver or mat knife between the papier-mâché and the mold to loosen the bowl before you pull it off the mold. Trim the top edges of the bowls with scissors to even them out.

If a student has papier-mâchéd over the rims of the 1/2 L container, you may need to cut the bowl right below the rims to get it off the mold.

Materials for Step 10
- *Screwdriver or mat knife*
- *Scissors*

11. Discuss the dried bowls

Ask students how the paper has changed after drying.

➤ *How did the newspaper strips change?*

➤ *Where did the water go that was in the wheat paste?*

12. Review vocabulary

When all students have engaged in the center activities, call them to the rug, and review the key vocabulary added to the word wall. Here's a suggested cloze review. Students answer chorally.

➤ *We _____ the paper to make it into strips.*

S: Tear.

➤ *The wheat paste is made from water and _____.*

S: **Flour**.

➤ *When we put strips on top of one another, we put on _____.*

S: Layers.

➤ *The technique we used for making things out of paper and paste is called _____.*

S: Papier-mâché.

across
around
dry
flour
layer
mold
over
papier-mâché
stiff
strip
wheat paste

> *When we have finished our papier-mâché bowls, we let them sit overnight to _____ .*

S: **Dry**.

> *When the paper and paste dries, it is _____ .*

S: **Stiff**.

Discuss these additional words and write them on the word wall.
- Across
- Around
- Mold
- Over
- Strip
- Wheat paste

13. Answer the focus question

Restate the focus question.

> *How can paper be made strong to form a bowl?*

Tell students you have a strip of paper with the question written on it. Review how to glue the strip into the notebook.

Ask students to answer the focus question in drawings and/or words.

WRAP-UP

14. Share notebook entries

Conclude Part 5 by having students share notebook entries. Ask students to open their science notebooks to the most recent entry. Read the focus question together as a class.

> *How can paper be made strong to form a bowl?*

Ask students to pair up with a partner to
- share their answers to the focus question;
- explain their drawings.

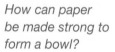

FOCUS CHART

How can paper be made strong to form a bowl?

We made a bowl from paper using papier-mâché.

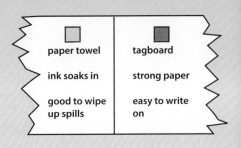

paper towel	tagboard
ink soaks in	strong paper
good to wipe up spills	easy to write on

INTERDISCIPLINARY EXTENSIONS

Language Extension

- **Make a paper chart**

  On the wall, display a large sheet of butcher paper Divide it into ten sections, and glue a paper sample at the top of each section. As students explore and discuss the different kinds of paper, add words and sentences to the chart under the appropriate paper sample. For example, facial tissue is good for blowing noses and wiping tears, but not very good for writing on. Put the list of uses for the different kinds of papers on a class chart.

Math and Engineering Extensions

- **Take apart and reassemble paper boxes**

  Have students take apart small boxes to see the shape of the piece used to form the box. Then have students try to reassemble the boxes. Use teacher master 21, *Math Extension—Paper Boxes*, to set this up as a center activity.

  Bring in a selection of small boxes made of chipboard—empty cereal boxes, paper-clip boxes, cracker boxes, toy boxes, and so on. (Milk cartons and other heavily coated boxes and two-part boxes do not work well for this activity.)

  Students should take a box apart along the seams where it has been glued together, being careful not to tear the box in other places. Their goal is to see what a box looked like when it was a flat piece of chipboard, before it was glued together to form a container. Demonstrate the procedure by taking a box apart. Have students help you determine which are the glued seams that should be taken apart.

  Tell students that after they take their boxes apart, they will slowly and carefully trace around the edge of the flattened box with a crayon or marking pen. Demonstrate tracing the outline of the box you took apart on a large sheet of newsprint or newspaper.

  After tracing, have students try to reassemble the boxes.

- **Make a paper box or paper envelope**

  Students can use the patterns on teacher master 22, *Math Extension—Make a Box*, and teacher master 23, *Math Extension— Make an Envelope*, to make useful objects from paper. Provide scissors and glue for these projects.

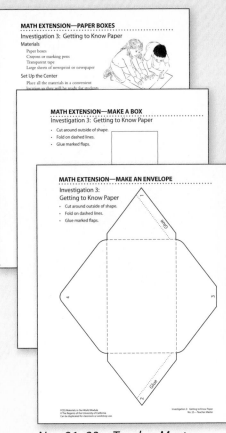

Nos. 21–23—Teacher Masters

Art and Engineering Extensions

- **Weave a paper mat**

  Have students weave paper mats. Refer to teacher master 24, *Art Extension—Paper Weaving*, when doing this activity at a center.

  Here's how to prepare the weaving materials and get students started.

  a. Cut one 15 × 23 cm (6" × 9") piece of construction paper for each student. Fold each piece in half, matching the short sides.

    Make five cuts in each paper, starting at the fold to about 2.5 cm (1") from the edge of the paper, every 2.5 cm. Do *not* cut all the way across.

    You can cut several bases at a time, using scissors or, better yet, a paper cutter. Fold the sheets in half, and place them with the folded side at the top of the paper cutter. Position and tape a small block of wood in such a way that it stops the blade after it has cut the proper distance into the paper.

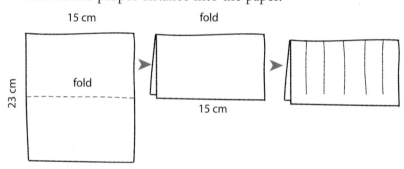

  b. Cut 2.5 × 23 cm (1" × 9") strips of construction paper to use for weaving. Use two colors (both different from the base color) so that each student will have three strips of each color (six strips in all). Cut extra strips for a demonstration.

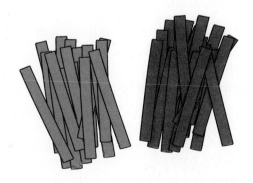

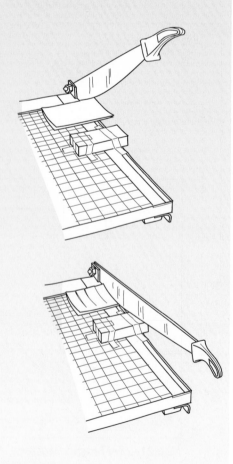

No. 24—Teacher Master

Give this demonstration to the class.

a. Tape the paper base to a flip chart or the board, so that the cuts go up and down.

b. Use one strip of paper to weave through the paper base. Have students chant with you as you demonstrate, "Under, over, under, over . . . ."

c. Push the woven strip down to the bottom of the paper, as far as it can go.

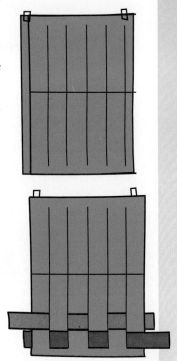

d. Use the same procedure to weave a second strip of another color. This is the difficult part for most early-childhood students. They must alternate the pattern, so wherever the strip went over before, it goes under this time, and vice versa. This time you start, "Over, then go under . . . ." Emphasize that with one color, students always start under (below or behind), and with the other color, students always start over (above or in front of).

e. Tell students to check the pattern after they finish weaving each strip. If they discover they've made a mistake, it's easy to pull the strip out and begin again. Show them how you can pull a strip out, then reweave it, with students telling you whether to go over or under.

f. Continue weaving strips until the entire paper is covered. This should take six strips, with a little room to spare.

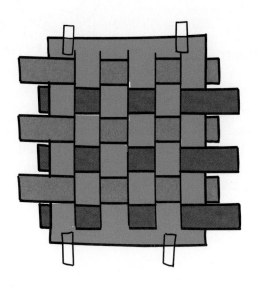

- ## Teach students simple origami

  Teacher master no. 25, *Art Extension—Pirate Hats*, outlines simple instructions for folding a paper hat from newspaper. This also makes an excellent in-class or at-home activity. *Math in Motion: Origami in the Classroom* by Barbara Pearl is a great resource for further projects.

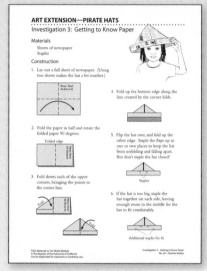

No. 25—Teacher Master

- ## Paint on paper

  Have students try out different kinds of paper on the paint easel. Ask them to find out which papers absorb the paint, which are too slick for the paint to stick to, and which might create different textures.

- ## Make collage masks

  Cut two eye holes in a paper plate. Glue a tongue depressor to the back of the plate to use as a handle. Let students decorate the mask, using a collage technique to cover the front of the paper plate.

- ## Examine paper-illustration techniques

  Torn or cut paper and collage techniques are sometimes used to produce illustrations in children's picture books. For example, Leo Lionni uses this technique in *Pezzettino*, Lois Ehlert uses it in *Red Leaf, Yellow Leaf*, and Steve Jenkins uses it in most of his books, such as *Biggest, Strongest, Fastest*. Discuss the illustrations in these books with students, and develop an art project based on these models.

- ## Construct a paper Humpty Dumpty
  Recite the rhyme "Humpty Dumpty" with students. Then make a paper Humpty Dumpty. Have students cut out oval shapes from 23 × 30 cm (9" × 12") sheets of construction paper. Have students glue on accordion-pleated or braided arms and legs and color or dress Humpty with paper clothes and a face.

Science Extensions

- ### Explore other kinds of paper
  Some additional kinds of paper that are interesting to study include crepe paper, carbon paper, and NCR (carbonless) paper. Look for discarded braille paper or other kinds of paper at your school. If students use carbon paper, they should be supervised to prevent damage to clothes and tables.

- ### Bring in rice paper
  Rice paper is made in some areas of the world for several purposes. Some rice paper is used for writing, and some is used for wrapping food. In the latter case, the wrapping is edible.

> **TEACHING NOTE**
>
> *Review the online activities for students on FOSSweb for module-specific science extensions.*

Home/School Connection

Students, working with adults, can make drinking cups for their families by following simple instructions.

Make copies of teacher masters 26 and 27, *Home/School Connection A and B* for Investigation 3, and send them home with students after Part 2.

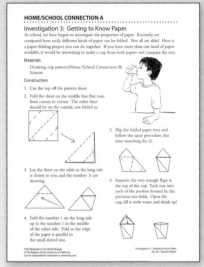

No. 26—Teacher Master

No. 27—Teacher Master

A second home/school connection describes making paper collages. This is a fun way to explore the many properties that different papers have.

Make copies of teacher master 28, *Home/School Connection C* for Investigation 3, and send them home with students.

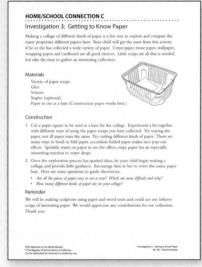

No. 28—Teacher Master

Investigation 4: Getting to Know Fabric

PURPOSE

Content

- Fabric is a flexible material with a wide range of properties.

- Fabric can be made of woven threads.

- Fabrics can absorb, transmit, or repel water.

- Wet fabric dries when water evaporates, leaving the fabric unchanged.

- Materials that interact with fabric may wash away or produce permanent stains.

- The properties of fabrics determine their uses.

Scientific Practices

- Observe and compare properties and structures of fabric.

- Observe and describe how and where fabrics are used.

- Observe how fabric interacts with water and other materials.

- Communicate observations made of different kinds of fabric, both orally and through drawings.

	Investigation Summary	Time	Focus Question
PART 1	**Feely Boxes and Fabric Hunt** Students observe the properties of ten different fabrics (burlap, corduroy, denim, seersucker, fleece, knit, ripstop nylon, sparkle organza, satin, and terry cloth). Students match properties by using feely boxes, hunting for fabric, and locating the fabrics that are used in the classroom.	**Center** 20 minutes **Fabric Hunt** 15 minutes **Notebook** 15 minutes	How are fabrics different? What is made of fabric?
PART 2	**Taking Fabric Apart** Students investigate the structure of woven fabrics by disassembling and comparing loosely woven burlap and tightly woven wool plaid.	**Center** 20 minutes **Notebook** 20 minutes **Reading** 20 minutes	How is fabric made?
PART 3	**Water and Fabric** Students conduct an investigation to find out how fabrics interact with water. Students discover the many ways the different fabrics absorb, transmit, and repel water. Students immerse fabric in water and observe that it is unchanged after it dries—the water evaporates.	**Center** 30 minutes **Notebook** 20 minutes **Follow-up (optional)** 15 minutes	What happens when water gets on fabric?
PART 4	**Soiling and Washing Fabric** Students go outdoors and soil a piece of muslin with dirt and plant materials. They attempt to clean the cloth, first by washing with plain water, then with detergent and a scrub brush. Students find that some substances make permanent stains.	**Introduction** 5 minutes **Center** 20–30 minutes **Notebook** 20 minutes	What are some things that stain fabric?
PART 5	**Graphing Fabric Uses** Students think about the kinds of fabric that would make a good pair of pants and other items of clothing. They prepare picture graphs that represent their decisions regarding the kinds of fabric they would use for different clothing applications.	**Introduction** 5–10 minutes **Graphing** 30–40 minutes **Reading** 20 minutes	How are different kinds of fabric used?

Content	Writing/ Reading	Assessment
• Fabric is a flexible material with a wide range of properties.	**Science Notebook Entry** Draw or write words to answer the focus questions.	**Embedded Assessment** Teacher observation
• Fabric can be made of woven threads.	**Science Notebook Entry** Draw or write words to answer the focus question. **Science Resources Book** "What Is Fabric Made From?"	**Embedded Assessment** Teacher observation
• Fabrics can absorb, transmit, or repel water. • Wet fabric dries when water evaporates, leaving the fabric unchanged.	**Science Notebook Entry** Draw or write words to answer the focus question.	**Embedded Assessment** Teacher observation
• Materials that interact with fabric may wash away or produce permanent stains.	**Science Notebook Entry** Draw or write words to answer the focus question.	**Embedded Assessment** Teacher observation
• The properties of fabrics determine their uses.	**Science Notebook Entry** Draw or write words to answer the focus question. **Science Resources Book** "How Are Fabrics Used?"	**Embedded Assessment** Teacher observation

BACKGROUND *for the Teacher*

Circus tents, T-shirts, washcloths, and 200-year-old Navajo rugs all have something in common. They are made of **fabrics**. Fabrics can be coarse like **burlap** or canvas, or fine like **satin**; thick like carpet or **terry cloth**, or thin like toile; loose like cheesecloth, or tight like **ripstop nylon**. In this module, the words *cloth* and *fabric* can be used interchangeably.

What Is Made of Fabric?

Fabrics are made from a wide variety of materials—from steel, brass, and other metals (hardware cloth or screen) to natural fibers (such as wool, silk, cotton, and hemp) to synthetic fibers (such as nylon, rubber, polyester, and polypropylene).

The properties of fabrics can be attributed in large part to the properties of the filaments or fibers from which they are made. Each different fiber produces a fabric with characteristic properties. Fabrics made from metal are extremely durable and long-lasting, but make very poor socks and furniture upholstery. Nylon, on the other hand, is light, flexible, and nearly waterproof but is not particularly good for warmth. Cotton fabrics absorb water quickly, making them good for towels and washcloths, but not so good for raincoats and tents.

Every fabric has properties that suit it for a particular purpose. The best fabric for a parachute is one of the worst fabrics for a carpet. The variety seems endless.

How Is Fabric Made?

A fabric is any material produced by **weaving**, knitting, or pressing fibers into a flexible sheet. **Denim**, burlap, and most other kinds of cloth are **woven**. In woven fabrics, one set of **threads** or yarns running in one direction is interwoven in some pattern with another set of threads or yarns running in a perpendicular direction. These two sets of threads are called the **warp** and **weft**. Woven fabrics can be identified by cutting a small piece and attempting to take it apart. If lots of little pieces of thread can be teased off the edges, the fabric is woven.

Knitted fabrics are produced from a single thread. Think about that sweater knitted by your aunt. She started with a big ball of yarn and turned it into a warm, spongy fabric with the aid of a couple of flying, clicking knitting needles. Knit fabrics stretch by the very nature of their manufacture and so are used for T-shirts, spandex, and socks.

Felt is made by mixing a variety of fibers, usually wool and fur for starters, and pressing the intermeshed fibers to make a hard, durable material. Unlike woven and knitted fabrics, felt tends to hold its shape because the fibers are locked together in a tight tangle. In the 19th century, the felting process used mercury, an insidious nervous-system poison when handled for extended periods of time. Hat makers often endured such exposure, resulting in the phrase "mad as a hatter."

What Happens When Water Gets on Fabric?

A thread of a natural fiber, such as cotton or silk, is made of hundreds of minute strands twisted together. The tiny spaces between the filaments provide locations for water to move by a process called capillary action. Fabrics that have this characteristic absorb water and many other liquids quite rapidly. Such fabrics take a longer time to dry because liquids can withdraw into tiny spaces to resist evaporation.

Fabrics made from larger, harder fibers, such as metals and plastics, have far fewer small spaces to hold water. Metal screens and plastic fabrics are poor for mopping up spills and don't allow effective movement of liquids through their structure by capillary action. Consequently, water dripped on a synthetic fabric will make a small wet spot, and additional water added to the spot will run through the fabric. Synthetic fabrics, such as nylon, polypropylene, and polyester, dry quickly after being soaked.

The rate of evaporation depends on four conditions: the amount of water vapor already in the air (humidity), the temperature, the amount of surface area exposed to the air, and air movement over the surface of the fabric. More evaporation occurs when the air is dry, when the temperature is high, when the surface area is large, and when air is moving rapidly over the surface of the cloth.

What Are Some Things That Stain Fabric?

We all struggle to keep the fabrics in our lives **clean**. Childhood is one constant assault on clothing with mud, grass stains, fruit juices, mustard, paint, chocolate, grease—you name it. When a soiling agent, such as

New Word

Say it → See it → Hear it → Write it →

Burlap
Clean
Cloth
Corduroy
Denim
Detergent
Dirty
Fabric
Fleece
Knit
Least
Most
Nubby
Ripstop nylon
Rough
Satin
Scratchy
Seersucker
Shiny
Slippery
Smooth
Soak
Soft
Sparkle organza
Stain
Terry cloth
Texture
Thread
Warp
Waterproof
Weaving
Weft
Woven

grape juice, falls on a fabric, a number of things can happen. First, the fabric might be **waterproof**, so the juice rolls off without leaving a mark. Usually, the grape juice **soaks** into the fabric and takes up space among the fibers of the threads where it becomes stuck in the mesh. The molecules in the grape juice might be chemically attached to the molecules in the fibers of the fabric in a process called **adhesion**. The pigments in the grape juice penetrate the individual fibers of the fabric and change their color.

If the soiling agent is water soluble, such as the sugar and other carbohydrates in grape juice, water will soften, dilute, and effectively remove the material. However, plain water is rarely sufficient to remove all the indiscretions of childhood; chemicals are usually needed. Soap and **detergent** are useful for removing soil that adheres tightly to fabrics. Soap acts to chemically change the nature of the adhering material, and detergent acts by making it possible for water to penetrate the junction between the soil and the fiber more effectively. Both are useful for removing organic material, oils, and a host of other stuff. But some materials actually change the color of the fibers in the fabric. Grape juice is quite good at turning cotton fibers purple. The color can't be washed out, and a **stain** results.

Stains can sometimes be removed with another chemical—bleach. Bleaches can chemically convert colored materials (pigments) to colorless ones.

How Are Different Kinds of Fabric Used?

Humans use fabric for clothing. Clothing provides warmth, protection, and status. Historically, important fabrics, such as cotton and flax, were woven from wools and plant fibers. These important fabrics were light, warm, **soft**, and flexible. For ceremonial occasions, the elegance of silk and lace declared the wealth and station of those who displayed them. Fabrics also cover our floors, windows, and furniture. These heavier fabrics are chosen for durability. Outdoor fabrics played roles in commerce and weather protection. Tough fibers woven into canvas provided sails for transportation and tents and other impromptu structures for protection against wind and storm. Today there are specialty fabrics performing precise functions in unique applications. One space-age fabric is used to repair defects in heart chambers and another to reinstate abdominal walls that have been compromised by hernia. Fabric is one of the most varied and versatile materials used by humankind.

TEACHING CHILDREN *about Fabric*

Early-childhood students are experts at getting fabrics wet and dirty. Students accomplish these feats almost without thinking about them. In this fourth investigation, students wet and soil fabrics intentionally, and, using their developing powers of observation and investigation, they will attend to the interactions between the fabrics and other materials.

The **conceptual flow** for Investigation 4 starts in Part 1 with the introduction of another class of **material—fabric**. Students observe the **properties** of ten different fabric samples, noting differences in **flexibility**, **softness**, **thickness**, and **texture**.

In Part 2, students study the **structure** of samples of burlap and wool plaid to learn how they are made. Students discover that fabric is made of **threads woven** together. Threads going one direction constitute the **warp**; the threads weaving through them constitute the **weft**.

In Part 3, students investigate the **interaction between fabric and water**. They first put drops on fabric and note that water **soaks** into some fabrics (absorbs), **drips** through some, and **beads** up on others. When students put fabric samples into a basin of water, the samples sink but otherwise are unchanged. When taken out of water, fabric **dries** when left exposed to air, and the water **evaporates**.

In Part 4, students go outdoors to test how fabric **interacts with other materials**. They **soil** fabric with dirt, vegetation, and, possibly, food. They then attempt to remove the soiling agents by **washing** with water, followed by washing with detergent. Any marks that remain are identified as **stains**.

In Part 5, students look at one of the most important **uses** of fabric, the manufacture of **clothing**. They consider several garments and make judgments about the fabric that would be most suitable for each. Using the knowledge they have been building throughout the module, students will be able to give specific reasons why some materials should be used for some kinds of clothing, but not for others. They display their thinking in bar graphs. These discussions will bring new light to students' observations of clothing and other uses for fabric they encounter in their daily lives.

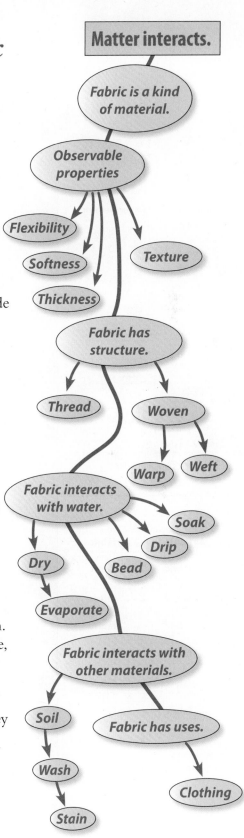

MATERIALS *for*
Part 1: *Feely Boxes and Fabric Hunt*

For each pair of students at the center

- 2 Fabric labels
- 2 Sets of five blue fabrics (See Step 3 of Getting Ready.)
  - Set 1 (burlap set): burlap, fleece, seersucker, organza, nylon
  - Set 2 (denim set): denim, corduroy, satin, knit, terry cloth
- 1 Cardboard box, approximately 30 × 25 × 18 cm (12" × 10" × 7") or brown grocery bag (See Step 4 of Getting Ready.) ★

For the class

- 70 Fabric squares (See Step 5 of Getting Ready.)
  - Burlap
  - Corduroy
  - Denim
  - Fleece
  - Knit
  - Ripstop nylon
  - Satin
  - Seersucker
  - Sparkle organza
  - Terry cloth
- 10 Zip bags, 1 L
- 1 Box cutter ★
- • Contact paper (optional) ★
- 1 Stapler (optional) ★
- • Transparent tape
- ❏ 1 Teacher master 3, *Focus Questions B*
- ❏ 1 Teacher master 29, *Center Instructions—Feely Boxes and Fabric Hunt*
- ❏ 1 Teacher master 30, *Fabric Labels*

For assessment

- ❏ • *Assessment Checklist*

★ Supplied by the teacher.   ❏ Use the duplication master to make copies.

CENTER INSTRUCTIONS—FEELY BOXES AND FABRIC HUNT

Materials
Sets of five blue fabrics
Feely boxes

Set Up the Center
Put a set of five fabrics at each student's place. Make sure both students in a pair have the same set of five fabrics. Keep the feely boxes close by, ready to use.

Guide the Investigation
1. **Explore sets of fabric.** When students are settled, ask them to explore the pieces of fabric in front of them. Guide their exploration by asking questions.
   - *How are the fabrics different?*
   - *How do they feel?*
   - *Describe the texture of two different fabrics.*
   - *Can you match your fabrics to your partner's? How?*
   - *Do you think you could match the fabrics if you couldn't see them?*
2. **Add to the word wall.** As students offer their observations, add new or important vocabulary to the word wall.
3. **Demonstrate the activity.** Put a set of five different fabric samples in a feely box and a matching set of samples on the table. Sit in front of the feely box (the side with no hole). Show students how to put their hands through the holes in the sides to feel the fabrics inside the box.
   Ask one student to choose a fabric from the set outside the box and put it in your hands, inside the box. Find the fabric that matches by feeling the pieces in the box. When you find a match, take both pieces of fabric out of the box and check. If the fabrics don't match, try again. If they do match, give one fabric back to the student helper, and put the other one back in the feely box. Let students begin.
4. **Switch places.** After the "feeling" students have matched all five of the fabric pieces, have students switch places. You can also switch sets of fabric so everyone gets to work with all ten fabrics.
5. **Wrap up the session.** Ask students to explain how they were able to tell the fabrics apart. Ask which fabrics were easy to tell apart and which were difficult.

Vocabulary
Try to include these words in discussions with students:
fabric, nubby, rough, scratchy, shiny, slippery, smooth, soft, sparkly, texture

No. 29—Teacher Master

FABRIC LABELS

This is made of **fabric**.	This is made of **fabric**.
This is made of **fabric**.	This is made of **fabric**.
This is made of **fabric**.	This is made of **fabric**.
This is made of **fabric**.	This is made of **fabric**.
This is made of **fabric**.	This is made of **fabric**.
This is made of **fabric**.	This is made of **fabric**.
This is made of **fabric**.	This is made of **fabric**.

No. 30—Teacher Master

FOCUS QUESTIONS B

Inv. 3, Part 1: **What is made of paper?**

Inv. 3, Part 2: **What makes paper good for writing? What makes paper easy to fold?**

Inv. 3, Part 3: **What happens when water gets on paper?**

Inv. 3, Part 4: **How can new paper be made from old paper?**

Inv. 3, Part 5: **How can paper be made strong to form a bowl?**

Inv. 4, Part 1: **What is made of fabric?**

Inv. 4, Part 2: **How is fabric made?**

Inv. 4, Part 3: **What happens when water gets on fabric?**

Inv. 4, Part 4: **What are some things that stain fabric?**

Inv. 4, Part 5: **How are different kinds of fabric used?**

No. 3—Teacher Master

GETTING READY *for*

Part 1: *Feely Boxes and Fabric Hunt*

1. Schedule the investigation

This part requires 15–20 minutes for the feely-box center for each group of six to ten students (3–5 pairs of students). Plan 5 minutes to introduce, 15 minutes for the whole-class fabric hunt after the work at the center, and 15 minutes for recording in notebooks and to wrap up.

2. Preview Part 1

Students observe the properties of ten different fabrics (burlap, corduroy, denim, seersucker, fleece, knit, ripstop nylon, sparkle organza, satin, and terry cloth). Students match properties by using feely boxes, hunting for fabric, and locating the fabrics that are used in the classroom. The focus questions are **How are fabrics different? What is made of fabric?**

3. Prepare the fabric samples for the feely boxes

Students work with sets of five fabrics in a feely box. Use the ten kinds of fabric samples to make two different sets.

Set 1 (burlap set) includes burlap, fleece, seersucker, sparkle organza, and ripstop nylon.

Set 2 (denim set) includes denim, corduroy, satin, knit, and terry cloth.

Once you know how many students will work at the center, make burlap sets for half the students and denim sets for the other half. Put the sets in separate zip bags. Each pair of students will work with the same set of fabrics for the feely-box activity.

4. Make feely boxes

Make a feely box for each pair of students working at the center. Gather medium-sized boxes, about 30 × 25 × 18 cm (12" × 10" × 7"). Cut holes in two opposite sides, large enough for students to insert their hands comfortably. Cut a third larger hole in the back, large enough for a second student to pass pieces of fabric into the box. Close the tops. You might want to cover the boxes with contact paper. Keep the boxes as permanent equipment in your classroom.

If boxes are scarce, compact feely bags can be constructed from brown-paper grocery bags (although these would be less durable). As with the boxes, cut two holes in the opposite sides for hands and a third hole in the back. Fold down the top of the bag, and staple it closed.

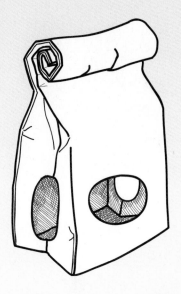

5. **Prepare the fabric for the hunt**
   After students work with the feely boxes, sort all the samples by fabric type. Select 30 pieces of fabric, three of each of the ten kinds of blue fabrics. While students are outside at recess, place the 30 pieces of fabric around the classroom where they can be easily found during a fabric hunt. Have 30 matching pieces of blue fabric ready to hand out to students.

   Have one or two extra pieces of each fabric ready for students who are unable to find a match. Sometimes mismatches or "hidden" pieces leave some students without a sample.

6. **Prepare labels**
   Make enough copies of teacher master 30, *Fabric Labels*, so that each student will have one label. Cut the labels apart ahead of time.

7. **Plan for tape distribution**
   Students will be using small pieces of tape to attach labels to objects made of fabric. Getting tape can cause a traffic jam. Put a number of little pieces of tape along the edge of a table or counter for students to get quickly during the activity.

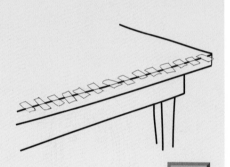

8. **Plan assessment**
   There are three objectives that can be assessed at any time during any part of this investigation.

   What to Look For

   - *Students ask questions (especially during the center discussions).*
   - *Students practice safety.*
   - *Students use tools appropriately.*

   Here are specific objectives to observe in this part.

   - *Fabrics have observable properties.*
   - *Fabrics can be compared and sorted by their properties.*
   - *Many objects are made from fabric.*
   - *Students communicate observations orally.*

   Focus on a few students each session.

9. **Collect wood, paper, and fabric scraps**
   The *Letter to Family* asked for contributions of wood, paper, and fabric scraps to be sent to school. Now is the time to send a reminder. Gather a variety of different kinds of paper that students will use on their sculptures in Investigation 5, Part 5—wrapping paper, magazines, crepe paper, wallpaper samples, and so on.

GUIDING *the Investigation*
Part 1: *Feely Boxes and Fabric Hunt*

1. **Introduce** *fabric*
   Call students to the rug. Tell them that for the next few days they will be studying **fabric**. Ask if they know what fabric is. Ask if they can name things they use or wear that are made of fabric.

2. **Focus question: How are fabrics different?**
   Write the first focus question on the chart, and read it together.

   ➤ *How are fabrics different?*

3. **Send students to the center**
   Tell students that they will have some time to look at several kinds of fabric at the center. Once they are familiar with the fabrics, they will do a matching activity. Send six to ten students to the center.

4. **Explore the sets of fabric**
   When students are settled at the center, ask them to explore the pieces of fabric in front of them. Guide the activity by asking,

   ➤ *How are the fabrics different?*

   ➤ *How do they feel?*

   ➤ *Describe the texture of two different fabrics.*

   ➤ *Can you match your fabrics to your partner's fabrics? How?*

   ➤ *Do you think you could match the fabrics if you couldn't see them?*

5. **Add vocabulary to the word wall**
   As students offer their observations, add any new or important vocabulary to the word wall. Acknowledge words students use, and offer new vocabulary as needed. Those new descriptive words might include, **nubby**, **rough**, **scratchy**, **shiny**, **slippery**, **smooth**, **soft**, **texture**.

6. **Demonstrate the feely-box activity**
   Put a set of five fabrics in a feely box and a matching set of fabrics on the table. Sit in front of the feely box (the side with no hole). Show students how to put their hands through the holes in the sides to feel the fabrics inside the box. Explain and demonstrate with a student helper as the "partner."

   a. *One student* (teacher in the demo) *selects a fabric from the set on the table outside the box and puts it in the partner's hands inside the box. The partner finds the fabric that matches by feeling each piece*

FOCUS QUESTION

How are fabrics different?
What is made of fabric?

TEACHING NOTE

Some students may know the term cloth. *Tell students that fabric and cloth are the same.*

Materials for Step 3
- *Sets of five fabrics*
- *Feely boxes*

TEACHING NOTE

Give students plenty of time to explore the fabrics before moving on to the feely-box activity.

already in the box. When the partner finds the match, the partner takes both pieces of fabric out of the box and checks them.

b. If the fabrics don't match, the partner tries again. If they do match, the partner puts one fabric back in the box and hands the other fabric to the other student.

7. **Start the feely-box activity**
Assign pairs of students to feely boxes. Each student in the pair uses an identical set of fabric samples. Let students start matching fabric samples. After the "feeling students" have matched all five fabric pieces, have students switch places so that everyone gets a chance at both roles in the activity. You can also switch sets of fabric so that all students get to work with all ten fabrics.

8. **Discuss the experience**
With the whole class together at the rug, ask students which fabrics were easiest to tell apart and which were hard to tell apart. Have a sample of each fabric for reference for the class. Give students time to describe other discoveries.

9. **Review vocabulary**
Review the key vocabulary words on the word wall so that students have words to describe the properties of the fabric. You might, at this time, introduce the word *cloth* which has the same meaning as *fabric*. If you feel that it might be confusing for students to have two words with the same meaning, save *cloth* for later.

cloth
fabric
nubby
rough
scratchy
shiny
slippery
smooth
soft
texture

B R E A K P O I N T

10. Introduce the fabric hunt

Tell students that you have placed pieces of ten kinds of fabric around the room. Explain that their task will be to look around the room and find one piece of fabric that matches the one that you will give them. When they find a fabric that matches, they should bring both pieces back to the rug, sit down, and lay the two pieces side by side. Emphasize that students should look for only *one* match and then return to the rug.

11. Begin the fabric hunt

Give each student one piece of fabric. Let students begin looking around the room for a fabric sample that matches the one they are holding. As they bring back pairs of fabric, check to make sure the fabrics are correctly matched. If they are not, help students identify the mismatch so they can try again.

Have extra samples ready for students to match if they are unable to find a match around the room.

12. Name the fabrics

When students are back on the rug, call for attention. Hold up a sample of each fabric, use a few words to describe its properties, say its name, and ask students to stand up when you name the kind of fabric they are holding. As you introduce the name, add it to the word wall.

- **Burlap**
- **Corduroy**
- **Denim**
- **Fleece**
- **Knit**
- **Ripstop nylon**
- **Satin**
- **Seersucker**
- **Sparkle organza**
- **Terry cloth**

13. Group students by kind of fabric

Tell students that their next challenge will be to find one or two other students who have the same kind of fabric. When you say "Go," students should stand up and move around the rug until they find another student who has the same kind of fabric. Tell them that they may find more than one partner. When all students have found their matches, have them sit together on the rug.

Materials for Step 10
- *Hidden fabric samples*
- *Matching fabric samples*
- *Extra fabric samples*

E L   N O T E

Model finding one sample of a matching fabric.

TEACHING NOTE

Some students may match satin with nylon. If they hold both fabrics up to the light, they will see a square pattern in the nylon, but not in the satin.

E L   N O T E

Tape a sample of each fabric next to its name on the word wall.

burlap
corduroy
denim
fleece
knit
ripstop nylon
satin
seersucker
sparkle organza
terry cloth

Materials for Step 16
- *Labels*
- *Transparent tape*

FOCUS CHART

What is made of fabric?

_____ is made of fabric.

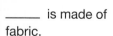

14. Review vocabulary

Review the vocabulary as you collect the fabric samples. Call out the name of one kind of fabric, and have students holding that kind of fabric come up and give you their pieces of fabric. Point to the name on the word wall as you call out the name of the fabric.

15. Focus question: What is made of fabric?

Write the second focus question on the chart, and read it together.

➤ *What is made of fabric?*

16. Search for fabric in the classroom

Tell students that there may be things made of fabric in the classroom. Give each student a "This is made of fabric." label and have students take a piece of tape to label something in the classroom that is made of fabric. When students are finished labeling, tell them to return to the rug.

17. Answer the focus question

Restate the second focus question.

➤ *What is made of fabric?*

Tell students you have a strip of paper with the question written on it. Review how to glue the strip into the notebook.

Ask students to answer the focus question in drawings and/or words.

WRAP-UP/WARM-UP

18. Share notebook entries

Conclude Part 1 or start Part 2 by having students share notebook entries. Ask students to open their science notebooks to the most recent entry. Read the focus question together as a class.

➤ *What is made of fabric?*

Ask students to pair up with a partner to

- share their answers to the focus question;
- explain their drawings.

MATERIALS *for*
Part 2: *Taking Fabric Apart*

For each student

1 Square of burlap, 4 cm (1.5")

1 Square of wool plaid, 4 cm (1.5")

2 Index cards

1 *FOSS Science Resources: Materials in Our World*

- "What Is Fabric Made From?"

For the class

1 Square of burlap, 4 cm (1.5")

1 Square of wool plaid, 4 cm (1.5")

1 Index card

- White glue ★

2 Loupes/magnifying lenses

❑ 1 Teacher master 31, *Center Instructions—Taking Fabric Apart*

1 Big book, *FOSS Science Resources: Materials in Our World*

For assessment

- *Assessment Checklist*

★ Supplied by the teacher.   ❑ Use the duplication master to make copies.

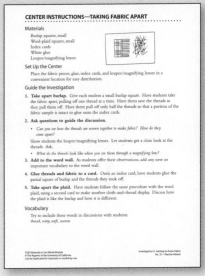

No. 31—Teacher Master

GETTING READY *for*
Part 2: *Taking Fabric Apart*

1. **Schedule the investigation**

   Plan 5–10 minutes to introduce this part to the whole class. Each group of six to ten students will need about 20 minutes at the center, or you can do this as a whole-class activity. Plan on 20 minutes for the reading and another 20 minutes for recording in notebooks and to wrap up.

2. **Preview Part 2**

   Students investigate the structure of woven fabrics by disassembling and comparing loosely woven burlap and tightly woven wool plaid. The focus question is **How is fabric made?**

3. **Plan to read** *Science Resources*: **"What Is Fabric Made From?"**

   Plan to read *"What Is Fabric Made From?"* in *Science Resources* during a reading period after completing the active investigation for this part.

4. **Plan assessment**

   Here is a specific objective to observe in this part.

   - *Fabrics are made from different materials.*

   Focus on a few students each session. Record the date and a + or − on the *Assessment Checklist*.

GUIDING *the Investigation*
Part 2: *Taking Fabric Apart*

FOCUS QUESTION

How is fabric made?

1. **Introduce the investigation**
   Call students to the rug. Show them the piece of burlap. Ask if they can tell what it is made of. [Lots of threads or strings put together.]

2. **Demonstrate taking fabric apart**
   Tell students that they will have a chance to take apart some pieces of fabric to see exactly how the fabrics were made. Demonstrate how to take threads off the piece of burlap until you are about halfway across the piece of fabric. Glue the remaining portion of the burlap sample on an index card, and then glue the loose threads on the same card next to the burlap.

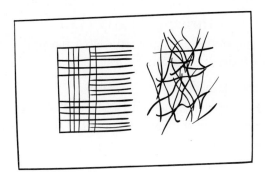

3. **Introduce *weft* and *warp***
   Tell students that some interesting words describe threads that make up fabric. Hold up the index card with the fabric and threads glued on it. Point out the threads that run crosswise on the fabric, and tell students,

   *These **threads** running side to side are called the **weft** of the fabric. The threads running up and down are called the **warp**.*

   Write these new words on the word wall and have students say them aloud.

4. **Focus question: How is fabric made?**
   Write the focus question on the chart, and read it together.

   ➤ *How is fabric made?*

TEACHING NOTE

Students enjoy using the words weft *and* warp *to talk about fabric; however, early-childhood students will not have a complete understanding of these words.*

Materials for Step 5
- *Burlap squares*
- *Wool-plaid squares*
- *Index cards*
- *White glue*
- *Loupes/magnifying lenses*

Say it
Write it
New Word
See it
Hear it

thread
warp
weft
woven

FOCUS CHART

How is fabric made?

Fabric is made from threads.

Fabric is made from woven threads.

5. **Take apart the burlap**
   Send six to ten students to the center. Give each student a small piece of burlap. Have students start to take the fabric apart, pulling off one thread at a time. Tell students to save the threads as they pull them. Have students pull off only half of the threads so that part of the fabric is intact to glue onto the card. Guide students' actions by asking,

   ➤ *Can you see how the threads are **woven** together to make fabric? How do they come apart?*

   Discuss what the fabric is made of and how the threads are woven. Show students the loupes/magnifying lenses and let them get a close look at the threads. Ask,

   ➤ *What do the threads look like when you see them through a magnifying lens?*

6. **Glue burlap and threads onto a card**
   Have students glue onto an index card the partial square of burlap and the threads students removed.

7. **Take apart the plaid**
   Have students use the wool plaid and a second index card to make another cloth-and-thread display. Discuss how the plaid is like the burlap and how it is different.

8. **Review vocabulary**
   When all students have engaged in the center activities, call them to the rug and review the key vocabulary added to the word wall.

9. **Answer the focus question**
   Restate the focus question.

   ➤ *How is fabric made?*

   Tell students you have a strip of paper with the question written on it. Review how to glue the strip into the notebook.

   Ask students to answer the focus question in drawings and/or words. Have students glue their index cards into their notebooks.

READING *in Science Resources*

10. Read "What Is Fabric Made From?"

Students have investigated different kinds of fabric and have observed how fabric is woven. This article increases students' awareness of the variety of natural resources used to make fabric.

Introduce the article. Ask students what fabric is made from. Brainstorm ideas, writing students' responses on chart paper. Have students listen for their ideas and for ideas that are new to them in the article.

Read the article aloud. Pause to discuss key points in the article, to review the pictures, and to make predictions.

11. Discuss the reading

Discuss the reading, using these questions as a guide.

➤ *What kind of fabric are you wearing? Where did it come from?*

➤ *Look closely at the fabric. How was it made? Is it woven or knitted?*

➤ *What is fabric made from?* [Animals' hair and fur, plants, oil.]

➤ *What should we take off our brainstorming list?*

12. Make a summary chart (optional)

Summarize the article by charting the resources used to make different fabrics. Make a two-column chart for the class with the headings "Fabric" and "What is it made from?" Prepare sentence strips by writing the words under the two headings as shown in the chart. Give students the prewritten sentence strips. Reread the article aloud or review it. Have students place their sentence strips on the class chart as you reread.

WRAP-UP/WARM-UP

13. Share notebook entries

Conclude Part 2 or start Part 3 by having students share notebook entries. Ask students to open their science notebooks to the most recent entry. Read the focus question together as a class.

➤ *How is fabric made?*

Ask students to pair up with a partner to

• share their answers to the focus question;

• explain their drawings.

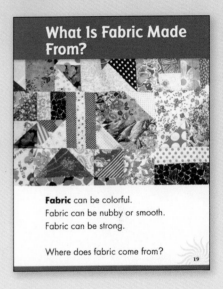

Fabric can be colorful.
Fabric can be nubby or smooth.
Fabric can be strong.

Where does fabric come from?

19

Fabric	What is it made from?
Wool	Sheep hair
Silk	Silkworm cocoons
Cotton	Cotton plants
Burlap	Jute plants
Nylon	Oil

MATERIALS *for*
Part 3: *Water and Fabric*

For each pair of students at the center

- 1   Set of ten blue fabrics (See Step 3 of Getting Ready.)
- 1   Plastic cup
- 2   Droppers

For the class

- 10   Zip bags, 1 L
- • Sponge
- • Paper towels ★
- • Clothesline and clothespins (optional) ★
- • Water ★
- • Newspaper (optional) ★
- 10   Containers, 1/2 L
- ❑ 1   Teacher master 32, *Center Instructions—Water and Fabric*

For assessment

- • *Assessment Checklist*

★ Supplied by the teacher.   ❑ Use the duplication master to make copies.

No. 32—Teacher Master

GETTING READY *for*
Part 3: *Water and Fabric*

1. **Schedule the investigation**
   This part requires 20 minutes at the center for each group of six to ten students, or it can be done as a whole-class activity with four students sharing a set of ten fabric samples. Plan 5–10 minutes to introduce and 20 minutes for recording in notebooks and to wrap up with the entire class at the rug.

2. **Preview Part 3**
   Students conduct an investigation to find out how fabrics interact with water. They discover the many ways the different fabrics absorb, transmit, and repel water. Students immerse fabric in water and observe that it is unchanged after it dries—the water evaporates. The focus question is **What happens when water gets on fabric?**

3. **Assemble fabric sets**
   Put one of each of the ten different blue fabric samples in a zip bag. Each pair of students at the center will get one set of these fabrics.

4. **Prepare the water**
   Fill the cups one-third full with water. You'll need one cup of water for each pair of students. Keep extra water on hand to refill cups if needed.

5. **Plan to dry the fabric pieces**
   At the end of the activity, hang the fabric samples on a classroom clothesline if you have one, or spread them out on paper to dry.

6. **Plan assessment**
   Here are specific objectives to observe in this part.

   - *Fabrics have observable properties.*

   - *Fabrics can be compared and sorted by their properties.*

   - *Many objects are made from fabric.*

   Focus on a few students each session. Record the date and a + or − on the *Assessment Checklist.*

What happens when water gets on fabric?

Materials for Step 3–6
- *Droppers*
- *Cups of water*
- *Pieces of fabric*
- *Paper towels*
- *Sponge*

Say it → See it → Hear it → Write it → **New Word**

GUIDING *the Investigation*
Part 3: *Water and Fabric*

1. **Introduce the investigation**
   Call students to the rug. Reintroduce the blue fabrics. Ask students what they think would happen if they put water on the fabrics. Tell students that they will be fabric scientists today as they investigate what happens when water gets on different fabrics.

2. **Focus question: What happens when water gets on fabric?**
   Write the focus question on the chart, and have students read it together.

   ➤ *What happens when water gets on fabric?*

3. **Drop water on terry cloth**
   Send six to ten students to the center. Make sure each pair of students has a set of the blue fabrics. Have students take the terry cloth out and lay it on the table. Ask them what they think will happen if they put a few drops of water on the cloth. Let students squeeze a few drops of water on the cloth. Discuss what students see.

4. **Add vocabulary to the word wall**
   As students offer their observations, add any new or important vocabulary to the word wall. Let students be the guides—acknowledge the words they use, and offer new vocabulary as needed. Some of the new words might be **least**, **most**, **soak**, and **waterproof**.

5. **Compare nylon and satin**
   Collect the terry cloth in the center of the table and dry off the tabletop with paper towels or a sponge. Have students take out the satin and nylon pieces, drop water on them, and observe what happens. Have students compare how the water acts on the different kinds of fabric.

6. **Observe all fabrics**
   Encourage students to drop water on the other pieces of fabric in the set, drying off the table between fabric tests. Continue to ask questions to guide students' investigations.

   ➤ *Does the water soak into the fabric? Bead up on top? Bead up first and then soak in?*

   ➤ *Which fabrics absorb the most water?*

➤ *How does the water move through the fabric? Does it stay in one spot or spread out?*

➤ *How does the fabric feel after you have one drop of water on it? Does it feel dry, or does it feel wet?*

7. **Discuss fabric uses**

Discuss the possible uses for fabrics that soak up water and those that are waterproof.

➤ *What kind of fabric would be good for a raincoat? A bath towel?*

8. **Hang the fabrics out to dry**

Hang the fabrics on the clothesline to dry. Use clothespins to hold them securely on the line.

9. **Review vocabulary**

When all students have engaged in the center activities, call them to the rug, and review the key vocabulary added to the word wall.

10. **Answer the focus question**

Restate the focus question.

➤ *What happens when water gets on fabric?*

Tell students you have a strip of paper with the question written on it. Review how to glue the strip into their notebooks.

Ask students to answer the focus question in drawings and/or words.

Materials for Step 8
- *Clothesline*
- *Clothespins*

least

most

soak

waterproof

FOCUS CHART

What happens when water gets on fabric?

On fabric, water sometimes _____.

WRAP-UP/WARM-UP

11. Share notebook entries

Conclude Part 3 or start Part 4 by having students share notebook entries. Ask students to open their science notebooks to the most recent entry. Read the focus question together as a class.

➤ *What happens when water gets on fabric?*

Ask students to pair up with a partner to

- share their answers to the focus question;
- explain their drawings.

BREAKPOINT

TEACHING NOTE

At this age, it is not expected that students will understand the mechanics of evaporation. This is a good opportunity for your young scientists to perceive the wonders of the everyday physical world, to take part in planning and conducting a simple investigation, and to gain firsthand experience with "drying up."

12. Investigate evaporation (optional)

Point out that the samples dried overnight. Ask,

➤ *What do you think happened to the water on the fabrics?*

➤ *Do you think the fabrics would dry in a plastic bag?*

➤ *How can we find out?*

Follow students' lead as you collaborate in planning a mini-investigation. Have a few students choose fabrics that absorb water, a few other students get the samples wet, and a few others place each sample in a 1/2 L container in a location they determine. Place similarly soaked fabric out in the open air for comparison. You might also suggest putting water in two containers—one with a lid and one without, and comparing them for a few days.

TEACHING NOTE

*See the **Home/School Connection** for Investigation 4 at the end of the Interdisciplinary Extensions section. This is a good time to send it home with students.*

MATERIALS *for*
Part 4: *Soiling and Washing Fabric*

For each student

1 Square of muslin, 15 cm (6")

For the class

6 Basins

1 Square of muslin

1 Permanent marking pen

• Craft sticks

• Staining materials ★

  • Mustard

  • Ketchup

  • Grape juice

  • Tempera paint

• Liquid laundry detergent or dish soap, 3/4 L (3 cups) ★

8 Squeeze bottles

8 Scrub brushes

1 Clothesline (or string) and clothespins ★

• Water ★

❑ 1 Teacher master 33, *Center Instructions—Soiling and Washing Fabric A*

❑ 1 Teacher master 34, *Center Instructions—Soiling and Washing Fabric B*

• *Outdoor Safety* poster

For assessment

• *Assessment Checklist*

★ Supplied by the teacher.   ❑ Use the duplication master to make copies.

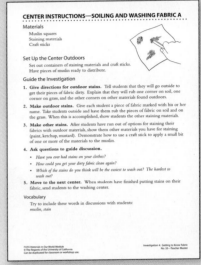

No. 33—Teacher Master

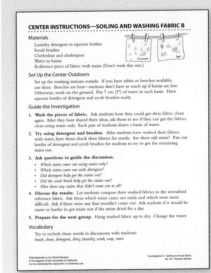

No. 34—Teacher Master

GETTING READY *for*
Part 4: *Soiling and Washing Fabric*

1. **Schedule the investigation**
   This part can be done as a whole-class activity or in centers. The outdoor activity will take 20–30 minutes outside. Plan on 5 minutes to introduce and 15 minutes for recording in notebooks. Allow 5 minutes to wrap up as you hang the fabric outside to dry.

2. **Preview Part 4**
   Students go outdoors and soil a piece of muslin with dirt and plant materials. They attempt to clean the cloth, first by washing with plain water, then with detergent and a scrub brush. Students find that some substances make permanent stains. The focus question is **What are some things that stain fabric?**

3. **Select your outdoor site**
   Search for a location with a variety of natural materials for staining—grass, soil, sand, mud, leaves, bark, puddles of water. Try to go out at a time of day when the grass is dry.

4. **Check the site**
   It is always a good idea to check the site the morning of an outdoor activity. Check for litter and debris and remove as necessary.

5. **Bring in helpers (optional)**
   Plan to have older students or adults help with the washing centers.

6. **Write students' names on muslin**
   Label a piece of muslin for each student, using a permanent marking pen. If you label in the lower right-hand corner, it will be easy to see the name when the muslin is hanging on the clothesline to dry.

7. **Make a sample**
   Stain a piece of muslin with all the staining materials. Try to keep each stain in a separate section of the cloth. Students will compare this unwashed sample to their pieces of cloth after washing.

8. **Put detergent in the bottles**
   Fill each squeeze bottle with liquid laundry detergent. You will need to fill the bottles just once for the entire class, because they dispense only a drop at a time.

   To fill the bottles, remove the dropper tip by pushing it sideways. The tip will pop off. After filling the bottle, push the tip back into place securely.

9. **Set up washing centers**

   Set up washing centers outside. If you have tables or benches available, use them. Benches are best—students don't have to reach up if the basins are low. Otherwise, work on the ground.

   Put 7 cm (3") of water in a basin for each pair of students. Have squeeze bottles of detergent and scrub brushes ready.

10. **Plan a place to dry the fabric**

    Ideally, you will hang the fabric flags near the paper flags from Investigation 3. If this is not possible, hang the fabric flags where they will be affected by wind, rain, and sun.

11. **Plan assessment**

    Here is a specific objective to observe in this part.

    - *Fabric can be changed by staining, washing, and drying.*

    Focus on a few students each session. Record the date and a + or − on the *Assessment Checklist*.

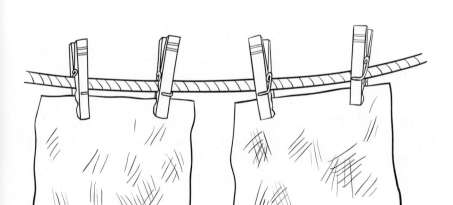

FOCUS QUESTION

What are some things that stain fabric?

Materials for Step 5
- *Squares of muslin*
- *Craft sticks*
- *Staining materials*

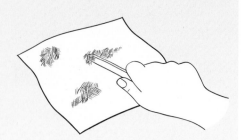

GUIDING *the Investigation*
Part 4: *Soiling and Washing Fabric*

1. **Introduce the investigation**

   Call students to the rug. Ask them to discuss how their clothes get dirty during the day, in the classroom and outdoors. Ask how clothes get clean after they are dirty. Tell students that, as fabric scientists, they will investigate how to get dirt out of fabric.

2. **Introduce the muslin**

   Show students a piece of muslin. Tell them that each student will get a piece that he or she can get dirty. After that, students will wash the muslin to see if they can get it clean again. Tell students that some of the dirt may not come out—the muslin might have some permanent **stains**.

3. **Focus question: What are some things that stain fabric?**

   Write the focus question on the chart, and read it together.

   ➤ *What are some things that stain fabric?*

4. **Give directions for outdoor stains**

   Tell students that they will go outside to get their piece of fabric **dirty**. Explain that they will rub one corner on soil, one corner on grass, and the other corners on other materials found outdoors. Students may suggest trying leaves, bark, flowers, fruit, and so on.

   You can use the *Outdoor Safety* poster to remind students of safe ways to conduct science investigations outdoors.

5. **Go outdoors**

   Give each student a piece of fabric marked with his or her name. Together, go to your home base outdoors in the usual orderly manner. Form a sharing circle and describe the limits of the activity area. Turn students loose to rub the cloth in the dirt and on the grass.

   After students have run out of options for staining their fabric with outdoor materials, show them other materials you have for staining (paint, ketchup, mustard). Demonstrate how to use a craft stick to apply a small bit of one or more of the materials to the muslin.

   After this is accomplished, call students back to the sharing circle.

6. **Ask questions to guide discussion**

   Ask,

   ➤ *Have you ever had stains on your clothes?*

   ➤ *How could you get your dirty fabric **clean** again?*

   ➤ *Which of the stains you made do you think will be the easiest to wash out? The hardest to wash out?*

7. **Wash the muslin with cold water**

   After students have shared their ideas, move to the washing stations. Ask students to see if they can get the cloth clean by using water only. Have students gather around the available basins.

8. **Use detergent and brushes**

   After students have washed the fabric with water, have them check the fabric for results. Ask students if there are still stains. Pass out bottles of **detergent** and scrub brushes for students to try to get the remaining stains out.

9. **Discuss the washing process**

   As students work at the basins, guide their observations by asking questions.

   ➤ *Which stains are coming out by using water only?*

   ➤ *Which stains are coming out by using detergent?*

   ➤ *Is detergent helping to get the stains out?*

   ➤ *Does the scrub brush help to get the stains out?*

   ➤ *Are there any stains that aren't coming out at all?*

10. **Discuss the results**

    Let students compare their washed fabrics to the unwashed reference fabric you prepared. Ask,

    ➤ *Which stains came out easily, and which were more difficult?*

    ➤ *Were there any stains that didn't come out?*

    ➤ *Would it be easier or harder to get the stains out of the fabric if you had allowed the stains to dry for a day? How could we find out?*

11. **Return to class**

    After students complete the washing, hang the pieces of muslin on an outdoor clothesline or indoors on a classroom clothesline. Return to the classroom.

Materials for Steps 7–8
- *Basins*
- *Bottles of detergent*
- *Scrub brushes*
- *Water*

Materials for Steps 10–11
- *Stained muslin piece*
- *Clothesline and clothespins*

> **TEACHING NOTE**
>
> *If possible, hang the muslin next to your paper flags outdoors. Students can compare the changes to the paper flags and the fabric flags when the flags are side by side. Students will see that fabric does not change as much as paper does during an equal time span.*

clean
detergent
dirty
stain

FOCUS CHART

What are some things that stain fabric?

Grass stains fabric.

_____ *stains fabric.*

---

12. Review vocabulary

Call students to the rug. Add new words to the word wall, and review key vocabulary that was introduced and used during the outdoor activity.

13. Answer the focus question

Restate the focus question.

➤ *What are some things that stain fabric?*

Tell students you have a strip of paper with the question written on it. Review how to glue the strip into the notebook.

Ask students to answer the focus question in drawings and/or words.

WRAP-UP/WARM-UP

14. Share notebook entries

Conclude Part 4 or start Part 5 by having students share notebook entries. Ask students to open their science notebooks to the most recent entry. Read the focus question together as a class.

➤ *What are some things that stain fabric?*

Ask students to pair up with a partner to

- share their answers to the focus question;
- explain their drawings.

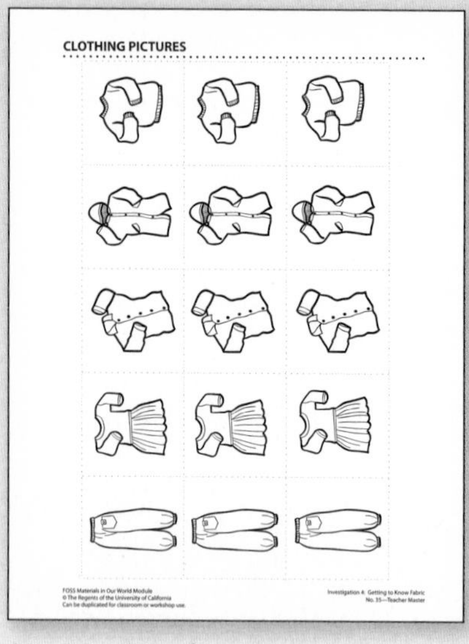

CLOTHING PICTURES

No. 35—Teacher Master

MATERIALS *for*
Part 5: *Graphing Fabric Uses*

For each student

- 1   *FOSS Science Resources: Materials in Our World*
  - • "How Are Fabrics Used?"

For the class

- • Chart paper (See Step 3 of Getting Ready.) ★
- • Sets of six blue fabrics (See Step 3 of Getting Ready.)
  - • Corduroy
  - • Denim
  - • Fleece
  - • Ripstop nylon
  - • Seersucker
  - • Sparkle organza
- • Masking tape
- • Transparent tape
- 1   Scissors ★
- ❏ 1   Teacher master 35, *Clothing Pictures*
- 1   Big book, *FOSS Science Resources: Materials in Our World*

For assessment

- • *Assessment Checklist*

★ Supplied by the teacher.   ❏ Use the duplication master to make copies.

GETTING READY *for*

Part 5: *Graphing Fabric Uses*

1. **Schedule the investigation**

   This part is a whole-class activity. Plan 30–40 minutes for the class. You may want to spread this activity over several days, graphing a different piece of clothing each day. Plan 5–10 minutes to introduce and 20 minutes for the reading.

2. **Preview Part 5**

   Students think about the kinds of fabric that would make a good pair of pants and other items of clothing. They prepare picture graphs that represent their decisions regarding the kinds of fabric they would use for different clothing applications. The focus question is **How are different kinds of fabric used?**

3. **Prepare fabric graphs**

   Make one (or more) large graph on chart paper. Draw lines, making the squares large enough for the clothing pictures (from teacher master 35, *Clothing Pictures*) to fit inside. You will need a separate graph for each article of clothing you choose to graph (e.g., one graph for pants, one graph for dresses, etc.). Consider laminating the graphs to make them a permanent addition to the kit.

   Tape six pieces of blue fabric at the bottom of each graph, one in each column: denim, corduroy, fleece, ripstop nylon, seersucker, and sparkle organza.

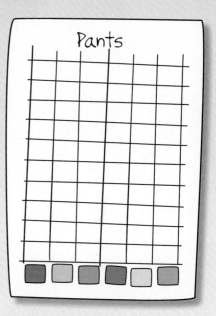

4. **Photocopy teacher master**

   Make copies of teacher master 35, *Clothing Pictures*. Cut the pictures apart, and group them by clothing type. Each student will need one picture of clothing for each type of graph you plan to create.

5. **Plan to read** *Science Resources*: **"How Are Fabrics Used?"**

   Plan to read "How Are Fabrics Used?" during a reading period after completing the active investigations for this part.

6. **Plan assessment**

   Here are specific objectives to observe in this part.

   - *Objects have properties that depend on the materials used to make them.*

   - *Students record and organize observations, using a graph.*

GUIDING *the Investigation*
Part 5: *Graphing Fabric Uses*

1. **Introduce the graph**

   Place the large graph on the rug. Call students to the rug and have them sit in a circle around the graph.

   Discuss the fact that different kinds of fabric are used for different purposes. For example, students found out that terry cloth soaked up the most water, and that is why it is used for towels, washcloths, and so on.

2. **Choose a cloth for making pants**

   Discuss the different kinds of fabric on the graph. Recall the names of the fabrics, and discuss a few of the properties of each. Discuss the properties that fabric should have to make a good pair of long pants (e.g., you can't see through it, it doesn't tear easily, it is warm).

3. **Describe the graphing procedure**

   Tell students that you are going to give them each a picture of a pair of long pants. When each student gets a picture, he or she should take it to the graph and put it in the column above the type of fabric he or she thinks would make the best pair of pants.

4. **Make a picture graph**

   Give each student a picture of the pair of pants. As soon as students have a picture, they should place the picture on the graph in the column above the fabric they think would make the best pants. They should place the picture in the lowest grid square available.

   You can tape the pants pictures on the graph with transparent tape.

5. **Ask questions about the graph**

   Ask students to look at the graph to answer these questions.

   ➤ *Which fabric was chosen most often for pants?*

   ➤ *How can you tell?* [That fabric has the most, or the longest line of, pants pictures.]

   ➤ *How many pants pictures are in each column?*

   ➤ *Were there any fabrics that were not chosen? How do you know?* [The graph has no pants pictures lined up above that fabric.]

   ➤ *Why was one fabric chosen so often for pants?* [The properties of the material make it a good choice.]

   Continue to ask questions, encouraging students to use the information provided by the graph.

Materials for Step 4
- *Chart-paper graph*
- *Pictures of pants*
- *Transparent tape*

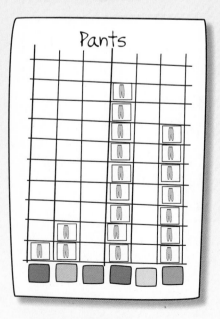

6. **Focus question: How are different kinds of fabric used?**
    Write the focus question on the chart, and read it together.

    ➤ *How are different kinds of fabric used?*

    Explain that students are going to make more graphs. Tell students that the graphs provide the information needed to answer the focus question.

7. **Make other graphs**
    Make a new graph for each additional item of clothing you would like students to graph. Follow the procedure described in Step 4 to make other graphs. There are additional pictures to graph students' fabric choices for shirts, dresses, jackets, and sweatshirts.

    Use a similar questioning strategy as in Step 5 to encourage students to use the information provided by the graph.

Materials for Step 7
- *Clothing pictures*
- *Chart-paper graph*
- *Transparent tape*

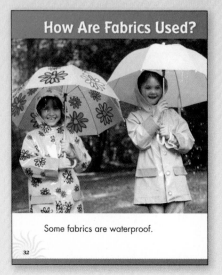

How Are Fabrics Used?

Some fabrics are waterproof.

32

FOCUS CHART

How are different kinds of fabric used?

_____ *fabric is used to make* _____.

READING *in Science Resources*

8. **Read "How Are Fabrics Used?"**
   This article expands students' knowledge by discussing how the characteristics of a fabric determine how to use it. For example, waterproof fabrics are good for outdoor uses, and fabrics that are easy to wash are good for clothing.

   Introduce the title of the article, "How Are Fabrics Used?" Have students listen for ideas in the article.

   Read the article aloud. Pause to discuss key points in the article, to review the pictures, and to make predictions.

9. **Discuss the reading**
   Discuss the reading, using these questions as a guide.

   ➤ *What kinds of fabric are you wearing?*

   ➤ *What kind of fabric is good for use outdoors? Why?*

   ➤ *What kind of fabric would you choose for a blanket? Why?*

10. **Answer the focus question**
    Restate the focus question.

    ➤ *How are different kinds of fabrics used?*

    Tell students you have a strip of paper with the question written on it. Review how to glue the strip into their notebooks. Ask students to answer the focus question in drawings and/or words.

INTERDISCIPLINARY EXTENSIONS

Language Extension

- **Make word and fabric cards**

  Make a set of cards for students to use to match the names of the fabrics to the samples. For each fabric, make three cards: a control card with a piece of fabric and its name, a card with just the fabric glued on, and a card with just the name.

  Use six to eight different fabrics for each set of cards. Have students lay out the control cards, then match the fabrics and the words to the control cards. After students have used the cards several times, challenge them to lay out the fabric cards, match the word cards to the fabrics, and then use the control cards to check their work.

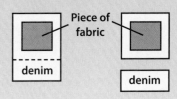

Piece of fabric

denim

denim

Math Extensions

- **Count seams**

  Ask students if their clothes are made of only one piece of fabric. Discuss seams as the places where pieces of fabric have been sewn together. Have students count how many seams can be found in shirts, pants, and so on.

- **Count pockets and make graphs**

  Make a bar graph. List pocket categories at the bottom of the graph, such as front pants pockets, back pants pockets, shirt pockets, and dress/skirt pockets. Have students use a self-stick note for each pocket they are wearing and stick the note in the appropriate column on the graph.

How Many Pockets?

| front pants pockets | back pants pockets | shirt pockets | dress/skirt pockets |

Bar graph

- **Make colorful caps**

  Have students use fabric paint or crayons to decorate cotton painters' hats. Read the classic story *Caps for Sale* by Esphyr Slobodkina. Follow up with math extensions such as counting the caps, putting a price on the caps (e.g., 10 cents), and finding out how much six caps would cost.

- **Make small pattern quilts**

  Use 2.5 cm (1") graph paper to make a quilt pattern. Cut 2.5 cm (1") squares of fabric for students to glue onto the sheets of graph paper, making a pattern of their choice.

TEACHING NOTE

Review the online activities for students on FOSSweb for module-specific science extensions.

Science and Engineering Extensions

- **Set up an ongoing fabric center**

  Set up a center where students can work independently and revisit activities such as feely boxes, collage, and weaving. You might also include other kinds of weaving activities, such as cardboard-lacing cards and weaving looms that use stretchy loops to make pot holders.

  Set up matching exercises for students. Have them match a variety of fabric pieces. Keep the size and shape of the pieces the same so students can concentrate on the fabric itself.

  Keep the loupes/magnifying lenses handy so students can look closely at the projects they are working on.

- **Show how knit fabric is made**

  Knit a small square of fabric, using yarn and knitting needles (or ask a parent or guardian to knit the square). Let students observe the square, possibly looking at it through a loupe/magnifying lens, and compare it to the pieces of blue knit and fleece fabric samples. When students are finished observing, unravel the square to show them how this kind of fabric is made with a continuous piece of yarn (or thread). Read *No Roses for Harry!* by Gene Zion after you have discussed the knitting process.

- **Simulate yarn-dye and piece-dye techniques**

  Obtain two pieces of fabric—a piece of wool plaid from Part 2 and a piece of fabric that has a print stamped on it. (You can tell this kind of material by looking at both sides; one side will be brightly colored, and the other will not.) You'll also need some muslin and fabric crayons or paint to simulate a piece dye.

  Show students the two kinds of material and discuss how they are alike and how they are different. Discuss the piece of weaving students did in the earlier activity. Compare it to the wool plaid. In this case, the threads used to make the fabric were dyed different colors before they were woven into the fabric.

  Have students look at the print fabric. This fabric was woven first, then printed with color. Give each student a piece of muslin, and let students draw a pattern, design, or picture on one side of the cloth with fabric crayons or paint to simulate the process used to make a print material.

- **Take a field trip**

  If possible, visit a farm where sheep shearing takes place or where cotton grows. Visit a clothing manufacturer to see how fabric is made into clothes or other products. Have students draw pictures showing the sequence from raw material, to fabric, to clothes.

TEACHING NOTE

Families can get more information about Home/School Connections from FOSSweb.

Home/School Connection

Students look around their homes for items that are made from fabric. They choose four items to illustrate on the student sheet. Families are encouraged to give clues: "I'm thinking of something that is made of fabric, and it is _____."

Make copies of teacher master 36, *Home/School Connection* for Investigation 4. Send it home with students after Part 3.

No. 36—Teacher Master

Investigation 5:
Earth Materials

INVESTIGATION 5 — Earth Materials

PURPOSE

Content

- Rocks, soil, and water are earth materials.

- Rocks can be compared, sorted, and described by their properties.

- Soil can be described by its properties.

- Water can change from solid to liquid (melt) with heat and from liquid to solid (freeze) with cold.

- Land, air, water, and trees are natural resources.

- People reuse and recycle to conserve natural resources.

Scientific Practices

- Observe and describe earth materials using the senses.

- Mix earth materials with water to observe properties.

- Use knowledge of the properties of materials to create useful and/or aesthetic objects.

Investigation Summary	Time	Focus Question
PART 1 — Exploring Earth Materials Students explore a collection of river rocks and organize them into groups based on size, shape, texture, and color. Students put the rocks in water and observe what happens. The class starts a rock collection at a rock center.	**Whole Class** 30 minutes **Notebook** 15 minutes **Reading** 20 minutes	How are rocks different?
PART 2 — Soil Painting Students collect soil samples from the schoolyard and mix the soil with water. The result is soil paint, which students use to create finger-paint art.	**Introduction** 10 minutes **Outdoors** 40 minutes **Notebook** 15 minutes **Reading** 15 minutes	What happens when soil gets wet?
PART 3 — Changes to Water Students freeze water in cups and observe how ice melts. Students learn that water can take different forms—liquid and solid—and can change from one form to another and back again.	**Whole Class** 15 minutes **Next Day** 25 minutes **Notebook** 15 minutes	How can we change water?
PART 4 — Reuse and Recycle Resources Students are introduced to natural resources and the need to reuse and recycle materials. They sort materials for recycling, based on the kind of material. Students use magnets to sort steel from other metals.	**Reading** 15 minutes **Recycling** 15 minutes **Notebook** 15 minutes	How can we conserve natural resources?
PART 5 — Making Sculptures Students fashion a sculpture, using wood, paper, fabric, glue, and natural materials. Students use the knowledge they have gained in the previous parts of the module to make their own creations.	**Introduction** 5–10 minutes **Center** 20–40 minutes **Reading** 15 minutes **Notebook** 15 minutes	What can we make from different materials?

At a Glance

Content	Writing/Reading	Assessment
• Rocks are earth materials. • Rocks have observable properties. • Rocks can be described, compared, and sorted by their properties.	**Science Notebook Entry** Draw or write words to answer the focus question. **Science Resources Book** "How Are Rocks Different?"	**Embedded Assessment** Teacher observation
• Soils are earth materials. • Soil can be described by its properties.	**Science Notebook Entry** Draw or write words to answer the focus question. **Science Resources Book** "How Are Rocks, Soil, and Water Used?"	**Embedded Assessment** Teacher observation
• Water is an earth material. • Water can change from solid to liquid (melt) with heat and from liquid to solid (freeze) with cold.	**Science Notebook Entry** Draw or write words to answer the focus question.	**Embedded Assessment** Teacher observation
• Land, air, water, and trees are natural resources. • People reuse and recycle to conserve natural resources.	**Science Notebook Entry** Draw or write words to answer the focus question. **Science Resources Book** "Land, Air, and Water"	**Embedded Assessment** Teacher observation
• People use knowledge of the properties of materials to create useful and/or aesthetic objects.	**Science Resources Book** "I Am Wood"	**Embedded Assessment** Teacher observation

BACKGROUND *for the Teacher*

Earth materials are any of the **solid**, **liquid**, or gaseous materials that make up Earth. **Rocks** are solid earth materials. They are relatively complex materials made of basic materials called minerals. Although there are a couple thousand minerals and many more kinds of rock, there are only about 25 common minerals and perhaps three times that many kinds of rock that could be considered common.

At one time or another, many children start rock collections. Some collections never extend beyond the pocket of a pair of jeans or a modest heap in the corner of a treasure box, but others take prominent places on windowsills or bookshelves at home. Rocks may capture young children's imaginations and spark an interest that endures for a lifetime.

What is it about rocks that makes them so collectible? They're durable, sometimes shiny, and occasionally sparkly. They come in different **sizes**, shapes, and colors, and they can look like jewels, especially when placed in water. You don't have to feed them, and they don't rot, dry out, or die if you forget to care for them. Beautiful, resilient, forgiving . . . what's not to like?

How Are Rocks Different?

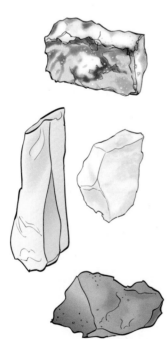

Geologists, the scientists who study Earth, are especially intrigued by the differences in rock composition. When geologists come across a rock that is new to them, they observe the physical properties of the rock and try to identify the components.

Observing rocks and beginning to sort them into **groups** are the initial steps children take in their role as geologists. Observing is a fundamental process in geology. Through observation, students gather information about rocks. Students use these observations to make comparisons and to sort the rocks into groups of similar properties. Students can also put the rocks in water to enhance the color and pattern.

Students often sort rocks by color first. With further challenges, students start to notice more subtle differences between rocks and begin sorting by properties such as shape, texture, weight, luster, and pattern. Even experienced geologists can surprise themselves when they observe the same rock a second time. Unexpected rock properties they hadn't noticed before jump out, ready to be recognized and appreciated.

What Happens When Soil Gets Wet?

What is **soil**? To the farmer, soil is the layer of earth material and organic detritus in which plants anchor their roots and from which they get the nutrients and water they need to grow. To a geologist, soil is the layer of earth materials at Earth's surface that has been produced by weathering of rocks and sediments. To an engineer, soil is any ground that can be dug up by earth-moving equipment and requires no blasting. To young students, soil is dirt.

In FOSS, *soil* is defined as a mixture of different-sized earth materials, such as gravel, sand, and silt. Soil also contains water and organic material called humus. Humus is the dark, musty-smelling stuff derived from the decomposed remains of plant and animal life. The proportions of these materials that make up soil differ from one location to another. Students will study this further in the **Pebbles, Sand, and Silt Module**.

Soil is a **natural resource**. All life depends on a dozen or so elements that are ultimately derived from Earth's crust. Soil has been called the bridge between earth material and life; only after minerals have been broken down and incorporated into the soil can plants process the nutrients and make them available to people and other animals.

Dry soil is lumpy or crumbly and falls apart easily in your hands. When mixed with water, soil changes. With a modest amount of water, soil holds together. You can mold a handful of moist soil into a ball. If you place it on a surface, there it is—a lump of moist soil.

If you add more water, you have **mud**. Mud is soil that is completely saturated with water. Mud is slippery and may flow and pour like a viscous liquid. All the spaces between the soil particles have been infiltrated by water. Mud may have the consistency of finger paint, and in this investigation, students use mud for artistic expression.

In the process of creating an earthy masterpiece, students attend to the texture of their paint and note the variety of colors that can be achieved with different soils and different dilutions of the raw material.

How Can We Change Water?

When conditions are right, samples of matter change state. They change from solid to liquid, liquid to gas, and solid to gas—or they change in the other direction. In this investigation, water is the subject of this transformation. Water in its solid state is called **ice**. When solid ice **melts**, it changes to liquid water. The water has not changed in terms of composition. Water is the same substance as ice. It is simply transformed into a different state.

When water evaporates, the water changes to gas. Water in its gaseous state is called vapor (and is invisible). Water vapor is made of the same particles as liquid water and ice. The three states of matter for any substance are composed of the same kinds of particles.

Change of state is directly related to temperature. You heat a solid to get it to melt. You heat it some more to get it to vaporize. The reverse is true as well. To progress from vapor to liquid to solid is a matter of cooling or **freezing** the substance. Students will start their understanding of this basic physical principle by putting liquid water in a freezer to see it turn to solid ice, and then watching the ice in class as it turns back to liquid water.

How Can We Conserve Natural Resources?

Conservation is a state of mind. It requires understanding the abstract ideas of interdependence, resource exploitation, and consequences. For young students, the first tentative steps toward an ethic to **conserve** may be learning some simple behaviors that honor and respect the practices of **recycling** and **reusing** the resources we extract from natural systems. The third *R* in the conservation mantra is reducing the amount of pressure we put on natural resources by using less. In time, we trust that our children will grow up to be conservation minded and prepared to tread more lightly on the earth that sustains us.

What Can We Make from Different Materials?

This question has no answer, or rather, it has an infinite number of answers. We are blessed on Earth with a bounty of natural raw materials and a staggering array of human-engineered and manufactured materials. Add to that the intangible ingredient of imagination, and anything is possible. The creativity that flows from the agile human mind can shape, bend, wrap, fold, heat, mix, layer, attach, and assemble materials to produce objects of utility, frivolity, and beauty to amaze and entertain us. Enjoy the show when your students create with scissors and glue.

Say it
New
Word
See it
Write it
Hear it

Conserve
Freeze
Group
Ice
Liquid
Magnet
Melt
Mud
Natural resources
Recycle
Reuse
Rock
Sculpture
Size
Soil
Solid

TEACHING CHILDREN *about* *Earth Materials*

In this last investigation, students have experiences with earth materials. Earth materials are often categorized as natural resources because of the central roles they play in civilization. It's not too early to start guiding students to an awareness of their place in the larger scheme of life on Earth and their responsibility for treading as gently as possibly as they pass through.

The **conceptual flow** for the investigation starts with a handful of samples of **rock**, one of the fundamental **earth materials**. In Part 1, students **observe** the **properties** of the rocks and sort them into a number of groups based on **shape**, **pattern**, **color**, **size**, and **texture**. Students discover that some properties are more easily observed when rocks interact with water.

In Part 2, students march out into the schoolyard with cups and spoons to collect samples of another key earth material, **soil**. They compare the soils from different locations in the schoolyard to note their **color** and **texture**. Students mix their soil samples with **water** to make "paint," with which they create artworks. They compare the color of the paints as another property of the soils.

In Part 3, students investigate a third earth material, **water**. They observe that **liquid** water placed in the **cold** turns into **solid ice** (**freezes**). When left in the **warm** classroom environment, the solid ice turns back into liquid water (**melts**).

In Part 4, the notion of **conservation of natural resources** is introduced. Students engage in an exercise of sorting discarded materials into categories based on several properties, with the idea that the materials could be **reused** or **recycled** to extend their use. The opportunity to introduce the idea of **reducing** use of resources comes up in the discussion of school and community recycling programs.

Part 5 is a **creative** activity for students. They are challenged to produce a sculpture from recycled materials.

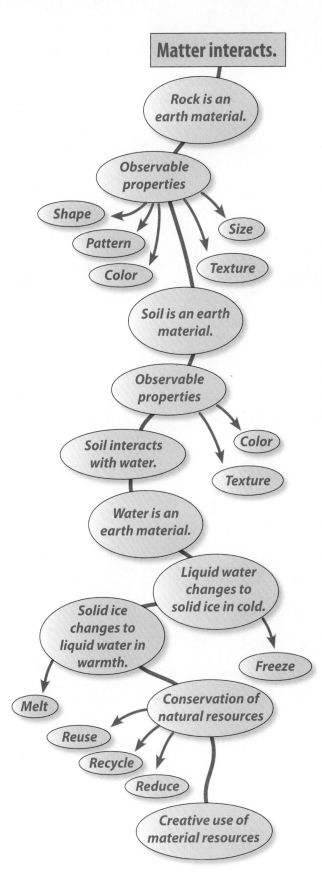

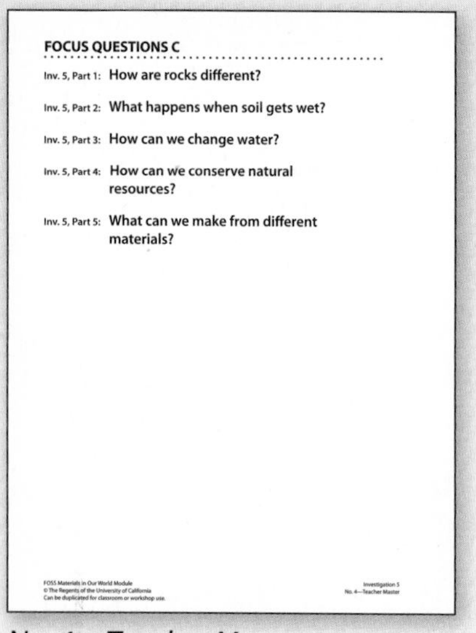

MATERIALS *for*

Part 1: *Exploring Earth Materials*

For each pair of students

1   Set of 20 large pebbles (See Step 3 of Getting Ready.)

1   Zip bag, 1 L

1   Plastic cup

2   Droppers

2   *FOSS Science Resources: Materials in Our World*

  • "How Are Rocks Different?"

For the class

1   Set of 20 large pebbles

1   Basin

1   Zip bag, 1 L

1   Balance (optional) ★

5   Loupes/magnifiying lenses

1   Pitcher or empty 2 L soft-drink bottle ★

1   Sponge

  • Paper towels ★

  • Water ★

1   Teacher master 4, *Focus Questions C*

1   Big book, *FOSS Science Resources: Materials in Our World*

For assessment

❏   • *Assessment Checklist*

★ Supplied by the teacher.   ❏ Use the duplication master to make copies.

FOCUS QUESTIONS C

Inv. 5, Part 1:   How are rocks different?

Inv. 5, Part 2:   What happens when soil gets wet?

Inv. 5, Part 3:   How can we change water?

Inv. 5, Part 4:   How can we conserve natural resources?

Inv. 5, Part 5:   What can we make from different materials?

FOSS Materials in Our World Module
© The Regents of the University of California
Can be duplicated for classroom or workshop use.

Investigation 5
No. 4—Teacher Master

No. 4—Teacher Master

GETTING READY *for*

Part 1: *Exploring Earth Materials*

1. **Schedule the investigation**
   This part is a whole-class activity and requires 25–30 minutes. Plan on 20 minutes for the reading and another 15 minutes for recording in notebooks and to wrap up.

2. **Preview Part 1**
   Students explore a collection of river rocks and organize them into groups based on size, shape, texture, and color. Students put the rocks in water and observe what happens. The class starts a rock collection at a rock center. The focus question is **How are rocks different?**

3. **Prepare river rock sets**
   Put 20 large pebbles (from the kit) in a zip bag for each pair of students. Make a set for yourself as well.

   If you have a rocky stream or beach close by, you might augment your classroom rock set by collecting 25–50 large pebbles (2–4 centimeters [cm] in diameter). Or you can take students to the schoolyard for a 10-minute rock hunt. Students are often surprised at how many rocks they can find around their school.

4. **Set up a rock center**
   Plan where you will display the rocks that students bring to class. Place the five loupes/magnifying lenses on the large, flat surface. This location can also serve as a rock-drying location.

5. **Plan to read *Science Resources*: "How Are Rocks Different?"**
   Plan to read "How Are Rocks Different" during a reading period after completing the active investigation for this part.

6. **Plan assessment**
   There are two objectives that can be assessed at any time during any part of this investigation.

   What to Look for

   - *Students ask questions.*

   - *Students communicate observations orally.*

   Here are specific objectives to observe in this part.

   - *Rocks have observable properties.*

   - *Rocks can be compared by their size, shape, color, and texture.*

   - *When rocks get wet, the colors are brighter.*

FOCUS QUESTION

How are rocks different?

Materials for Step 3
- *Bags of large pebbles*
- *Loupes/magnifying lenses*

Materials for Step 5
- *Cups of water*
- *Droppers*
- *Paper towels*

GUIDING *the Investigation*
Part 1: *Exploring Earth Materials*

1. **Introduce river rocks**

   Call the class to the rug. Hold up your bag of river rocks. Show students some of the rocks and tell them,

   *Today I brought some **rocks** to class. They are the kind of rocks you might find in a stream or along the side of a river.*

2. **Focus question: How are rocks different?**

   Write the focus question on the chart, and read it together.

   ➤ *How are rocks different?*

   Tell the class that they are going to explore the river rocks to find out how they are different from one another.

3. **Introduce sorting**

   Each pair of students will get a bag of rocks. Tell students they will share the rocks and observe them closely with the loupe/magnifying lens. When they find rocks that look or feel the same, students can put the rocks together in a group.

   Distribute a bag of rocks to each pair and have students begin exploring at their tables.

4. **Monitor the sorting**

   Monitor the pairs as they observe and sort the rocks into groups. Ask questions to stimulate their thinking.

   ➤ *Why did you put these rocks together?*

   ➤ *Can you **group** the rocks by color or by pattern?*

   ➤ *Are all the rocks the same **size**? How many size groups can you make?*

   ➤ *How do the rocks feel (texture)? Put rocks that have the same texture in a group.*

   ➤ *Group the rocks by shape. Describe the shape in each group.*

5. **Put drops of water on rocks**

   Call for students' attention and ask,

   ➤ *What might happen if you put drops of water on the rocks?*

   Distribute a cup of water, two droppers, and a paper towel to each pair of students. Describe the procedure.

   a. *Tear the paper towel in half.*

   b. *Put a rock on a piece of paper towel.*

   c. *Put drops of water on the rock and observe.*

After several minutes of dropping water on rocks, ask,

➤ *How did the rocks change?* [Colors got brighter; patterns appear; the rocks are shiny, sparkly, and darker.]

6. **Put the rocks in water**
   Suggest that students put one rock in the cup of water. Ask,

   ➤ *What happened to the rock?* [Sunk, got darker, got shiny.]

   ➤ *What happened to the water?* [Got cloudy or muddy.]

   Let students put additional rocks in the water and observe.

7. **Re-sort the wet rocks**
   Ask students to work with their partners to sort the wet rocks.

8. **Clean up**
   Find a place where students can set the rocks out to dry. Have students pour the remaining water into a basin at the materials station and wipe up any spilled water. Collect the plastic cups. Reuse or recycle the gray water to water some plants.

9. **Add vocabulary to the word wall**
   Gather students at the rug, and ask them to add new words about rocks to the word wall. Let students be the guides—acknowledge the words they use, and offer new vocabulary as needed.

group
rock
size

10. **Answer the focus question**
    Restate the focus question.

    ➤ *How are rocks different?*

    Distribute a strip of paper with the focus question for students to glue into their notebooks. Ask students to answer the focus question in drawings and/or words.

WRAP-UP/WARM-UP

11. **Share notebook entries**
    Conclude this session by having students share notebook entries. Ask students to open their science notebooks to the most recent entry. Read the focus questions together as a class.

    ➤ *How are rocks different?*

    Ask students to pair up with a partner to

    - share their answers to the focus question;
    - explain their drawings.

> **FOCUS CHART**
>
> *How are rocks different?*
>
> *Rocks have different patterns.*
>
> *Rocks are different sizes.*
>
> *Rocks are _____ .*

How Are Rocks Different?

Think about a rock.
What are its **properties**?
What does the rock look like?

41

READING *in Science Resources*

12. Read "How Are Rocks Different?"

Introduce the article "How Are Rocks Different?" Ask students to share briefly what they have learned about rocks. Continue the discussion by asking if students think that what they have learned will be in this article. Ask them to listen for their ideas or ones that are new to them.

Read the article aloud. This article asks students to think about a rock. Pause to allow them to develop an image of a rock in their minds.

13. Discuss the reading

Discuss the reading by asking students to work with a partner.

➤ *Turn to your partner. Describe to your partner the properties of the rock you were thinking about.*

Later, students may want to draw pictures of their rocks and write or dictate sentences to an adult describing their rocks.

14. Start a class rock collection

Ask students to think about where they should look for rocks. Record their answers on a sheet of chart paper. Title the chart "Where to Look for Rocks."

Ask students to volunteer any rules they think they should follow when collecting rocks outdoors. For example, the rocks should be smaller than your fist; only take two rocks; have an adult with you.

Show students where they can display the rocks they bring to class. Have cups of water, droppers, loupes/magnifying lenses, and a balance (optional) on the table for students to use.

15. Read about special rocks (optional)

Read the classic book *Everybody Needs a Rock* by Byrd Baylor. The book suggests ten rules for finding a special rock. A number of the rules deal with features such as size, shape, color, and texture. Ask students if they agree or disagree with the rules. Discuss how or why students might change the rules.

TEACHING NOTE

Refer to the online teacher resources on FOSSweb for a list of appropriate trade books that relate to this module.

MATERIALS *for*
Part 2: *Soil Painting*

For each pair of students

- 1 Container, 1/2 L
- 1 Metal spoon
- 2 Paper plates or sheets of white construction paper (28 × 45 cm; 11"× 18") ★
- 2 Clipboards (Optional; see Step 5 of Getting Ready.) ★
- • Soil (See Step 3 of Getting Ready.) ★
- 2 *FOSS Science Resources: Materials in Our World*
  - • "How Are Rocks, Soil, and Water Used?"

For the class

- 2 Basins
- 2 Zip bags, 1 L
- • Soil samples (See Step 4 of Getting Ready.) ★
- 1 Garden trowel (optional) ★
- • Water ★
- 1 Pitcher or empty 2 L soft-drink bottle ★
- 1 Permanent marking pen
- 1 Cloth towel ★
- 1 Big book, *FOSS Science Resources: Materials in Our World*

For assessment

- • *Assessment Checklist*

★ Supplied by the teacher.

GETTING READY *for*

Part 2: *Soil Painting*

1. **Schedule the investigation**

   This outdoor session will take about 40 minutes. Plan on 10 minutes to introduce the session indoors. In addition, plan 15 minutes for the reading and 15 minutes for recording in notebooks and to wrap up.

2. **Preview Part 2**

   Students collect soil samples from the schoolyard and mix the soil with water. The result is soil paint, which students use to create finger-paint art. The focus question is **What happens when soil gets wet?**

3. **Select your outdoor site**

   Find a location where students can dig up some soil without disturbing the landscaping and a second location where they can do their soil painting. When finished, students can return the "paint" to the location where they originally collected it.

   Check your site on the morning of the activity to make sure that everything is suitable for the class. If the weather is wet or windy, plan on going out another day.

4. **Gather soil samples**

   Gather two soil samples of various colors to show the class during the introduction. Place them in two 1 L zip bags.

   If you are teaching this module during the winter when the ground is frozen, you may need a garden trowel to acquire soil samples for the activity.

5. **Select art paper**

   Decide what kind of paper you want students to use for soil paintings. White construction paper or paper plates work well. Students enjoy the space provided by a 28 × 45 cm (11" × 18") sheet of construction paper.

   Write students' names on each paper (or plate) with a permanent marking pen. Carry the papers outdoors and distribute them after students have prepared their soil paint.

   If the day is breezy, think about how to anchor the paper. Clipboards are good, as are a number of impromptu paperweights. The 1/2 L container of soil paint can help hold paper, as can a rock or other weighty object.

6. **Plan to store artwork flat**

   Plan to store students' paintings in a place in the classroom where the paintings can lay flat to dry.

7. **Consider turning soil into paint**

   You might want to practice this procedure ahead of time. Students gather soil in a 1/2 L container (about one-quarter full), using a metal spoon. After discarding anything that is not soil, students pulverize the soil with their spoons to break down the clods, and then they add water. A good ratio is about four parts soil to one part water. Students mix the soil paint with a spoon and use their fingers as a brush.

8. **Plan for washing hands**

   Provide two basins half filled with water for hand washing. Students will rinse their hands in the first basin and then again in the second basin. Provide a cloth towel for students to dry their hands.

   After students have rinsed their hands in the basins, rinse the 1/2 L containers in the dirty water and reuse the water to water some plants.

   Students should wash their hands thoroughly with soap and warm water back in the school building. The containers can be rinsed again when back in the classroom.

9. **Collect materials for recycling in Part 4**

   Begin to collect materials that can be reused or recycled in Part 4. Try to obtain steel and aluminum cans so students can use magnets to separate them. Also bring in a variety of different plastic containers, some glass bottles, and an array of paper products, including newspapers and cardboard boxes.

10. **Plan to read *Science Resources*: "How Are Rocks, Soil, and Water Used?"**

    Plan to read "How Are Rocks, Soil, and Water Used?" during a reading period after completing the outdoor activity.

11. **Plan assessment**

    Here are specific objectives to observe in this part.

    - *Soil is an earth material.*

    - *Soil has small rocks it in.*

    - *A mixture of soil and water can be used as paint.*

    Focus on a few students each session. Record the date and a + or − on the *Assessment Checklist.*

FOCUS QUESTION
What happens when soil gets wet?

Say it
Write it
New Word
See it
Hear it

Materials for Step 3
- *Soil samples*
- *Container*
- *Spoon*

Materials for Step 4
- *Paper*
- *Metal spoons*
- *Containers*
- *Pitcher of water*
- *Basins of water*
- *Cloth towel*

GUIDING *the Investigation*
Part 2: *Soil Painting*

1. **Introduce the investigation**
   Call students to the rug. Review washing rocks. Ask,

   ➤ *When we put our rocks in water, what happened to the rocks?*

   ➤ *What happened to the water?*

   Tell students,

   Today we are going to add water to **soil**.

2. **Focus question: What happens when soil gets wet?**
   Write the focus question on the chart, and read it together.

   ➤ *What happens when soil gets wet?*

   Tell the class that they are going to mix soil and water to find out what happens when soil gets wet. Explain,

   We are going to collect soil in a container, add water, and mix it with a spoon. Then we are going to use the mixture and our fingers to paint on paper.

3. **Demonstrate the procedure**
   Hold up the two soil samples you collected. Pass them around for students to observe. Ask,

   ➤ *Are the two samples the same?* [No, different color, possibly different texture.]

   Tell students that they will collect different soil samples from the schoolyard. Describe and demonstrate this procedure.

   a. *You and your partner will have a cup like this* (1/2 L container).

   b. *Use a spoon to dig up the soil. Each student can put four spoonfuls of soil in the cup.*

   c. *Take out all the leaves, sticks, and rocks.*

   d. *Break the soil into small pieces.*

4. **Go outdoors**
   Have students line up in pairs. Distribute a 1/2 L container to one student and a spoon to the other student in each pair. Walk to your home-base location in the usual orderly manner.

   Define the limits of the collection area. Let students collect soil.

5. **Set up for soil painting**

   After each pair of students has collected 8 spoonfuls of soil, lead them to the painting location with their containers of soil. When the soil is free of debris and adequately processed by breaking the clods, add water to each container (one part water to four parts soil). After students stir with the spoon, ask,

   ➤ *You put soil and water together to make a mixture. How did the soil change when you made the mixture?* [The soil got wet and turned into **mud**.]

   ➤ *How does it feel?* [Wet, slippery, cool, smooth.]

6. **Begin soil painting**

   Distribute paper to each student, and have students create a design with the soil paint by using their fingers and hands. Ask,

   ➤ *How does the soil paint look on the paper?*

   ➤ *How does the soil and water mixture feel on your fingers? What is the texture of the mixture?*

   ➤ *What does the soil paint look like when it is wet?*

   As students finish painting, have them place their art in the sunshine to dry. They should scrape the unused soil paint back onto the ground where it came from. Have them place their empty containers near the wash basins for cleaning.

7. **Clean up**

   Have students rinse their hands in the first basin of water and then, again, in the second basin. Once all hands are rinsed, ask a few students to rinse the 1/2 L containers. Pour the water out of the basins to water some plants, and use the basins to carry the spoons and 1/2 L containers inside.

8. **Return to class**

   Demonstrate how to carry the art inside, keeping it level so the soil paint won't drip. Lay the paintings flat in the predetermined spot.

9. **Review vocabulary**

   Add new words to the word wall, and review vocabulary that was introduced and used during the outdoor activity.

10. **Answer the focus question**

    Restate the focus question.

    ➤ *What happens when soil gets wet?*

    Distribute a strip of paper with the focus question for students to glue into their notebooks. Ask students to answer the focus question in drawings and/or words.

New Word — Say it, See it, Hear it, Write it

 mud
soil

FOCUS CHART

What happens when soil gets wet?

Water and soil make _____.

READING *in Science Resources*

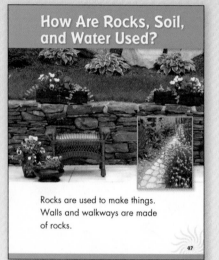

How Are Rocks, Soil, and Water Used?

Rocks are used to make things. Walls and walkways are made of rocks.

47

11. Read "How Are Rocks, Soil, and Water Used?"

Introduce the title, "How Are Rocks, Soil, and Water Used?" Ask students to list briefly some ideas they have now. As you read to students, ask them to listen for those ideas. Ask students to also listen for new ideas.

Read the article aloud. Pause to discuss each resource.

12. Discuss the reading

Discuss the reading, using this question as a guide.

➤ *What new ideas did you hear in the article about the use of rocks, soil, or water?*

WRAP-UP/WARM-UP

13. Share notebook entries

Conclude Part 2 or start Part 3 by having students share notebook entries. Ask students to open their science notebooks to the most recent entry. Read the focus question together as a class.

➤ *What happens when soil gets wet?*

Ask students to pair up with a partner to

- share their answers to the focus question;
- explain their drawings.

14. Display the artwork

Create a bulletin board displaying students' soil paintings.

MATERIALS *for*
Part 3: *Changes to Water*

For the class

- 7   Plastic cups
- 6   Craft sticks
- •   Ice chest or cooler (optional) ★
- 1   Basin
- 1   Pitcher or empty 2 L soft-drink bottle ★
- •   Soil sample ★
- •   Rocks (optional)

For assessment

- •   *Assessment Checklist*

★ Supplied by the teacher.

GETTING READY *for*

Part 3: *Changes to Water*

1. **Schedule the investigation**
    This part is a whole-class investigation. Plan 15 minutes for the class to set up the investigation on the first day and 25 minutes the second day to observe the results. Plan 15 minutes for recording in notebooks and to wrap up.

2. **Preview Part 3**
    Students freeze water in cups and observe how ice melts. Students learn that water can take different forms—liquid and solid—and can change from one form to another and back again. The focus question is **How can we change water?**

3. **Prepare for freezing**
    You will need space to put six plastic cups with water in a freezer overnight. Using a freezer at school is best because students can actually participate in putting the cups in the freezer and taking them out.

    Check for available space if you are considering using the freezer in the teachers' room. Check with food service if you are considering using the freezer in the cafeteria.

    If a freezer at school is not available, you can freeze the cups of water at home and bring them to school in an ice chest or cooler.

4. **Plan assessment**
    Here are specific objectives to observe in this part.

    *   *Water can be a liquid or a solid.*

    *   *When liquid water freezes, it turns to a solid—ice.*

    *   *When solid ice melts, it turns to a liquid—water.*

    Focus on a few students each session. Record the date and a + or − on the *Assessment Checklist*.

GUIDING *the Investigation*
Part 3: *Changes to Water*

1. **Introduce a cup of water**
   Call students to the rug. Pour about 150 milliliters (mL) of water from a pitcher into each of two cups. Ask students to describe some of the properties of water. [Liquid, flows, pours, wet, clear.]

2. **Focus question: How can we change water?**
   Write the focus question on the chart, and read it together.

   ➤ *How can we change water?*

   Ask students for their ideas on how we can change water.

   Students might suggest putting something in the water such as rocks or soil as they did in the previous part. Add some soil to one of the cups of water, and ask students what is different. They will see a color change, and the water may not be as clear as it was before.

   Students might suggest heating or cooling the water. Agree with their idea.

3. **Discuss putting water in a freezer**
   Ask students,

   ➤ *What will happen to this cup of water if we put it in a freezer? Will it stay the same, or will it change?* [It will **freeze**, make ice.]

   Write students' ideas on the board or the focus chart.

4. **Set up cups of water**
   Tell students,

   Six cups of water just like this one will go into the freezer. This cup of water will stay here in the classroom for comparison.

   With the help of some volunteers, set up six additional cups.

   a. Pour about 150 mL of water into six cups.

   b. Insert a craft stick into each cup.

   c. Place the six cups in a basin for transport to the freezer.

5. **Take a field trip to the freezer**
   Either as a class or in groups, go to the freezer at the school and place the cups in the freezer, where they will remain overnight. The investigation will continue the next day.

Materials for Steps 1–2
- *Pitcher of water*
- *Cups*
- *Soil sample*
- *Rock (optional)*

Say it
Hear it
See it
Write it
New Word

Materials for Step 4
- *Cups with water*
- *Craft sticks*
- *Basin*

B R E A K P O I N T

6. **Remove the water from the freezer**
   Take a second field trip to the freezer to retrieve the cups. Back in the classroom, put them on a table next to the cup of liquid water for comparison. Call the class around. Lift the ice on a stick from one of the cups to confirm that the water is ice. Place the ice back in the cup.

7. **Describe the changes**
   Ask students to describe how the water changed. Confirm that

   *The water that we drink and use to water plants is **liquid** water. Liquid water pours and splashes.*

   *The water we took out of the freezer is **solid** water called **ice**. Solid water (ice) is hard, and it stays in one shape.*

   In the freezer, liquid water changed into solid water.

8. **Observe the ice over time**
   Ask students to predict what will happen to the ice that is sitting out in the classroom. Write their ideas on the focus chart.

   Tell students that the ice will remain at the tables, and during the day they will make observations to see what happens. Plan on making observations every 30 minutes or so.

   In addition to looking at the change, students can touch the water and the ice as it **melts**. As students describe the changes, write their descriptive words on the word wall. These are some of the things students will see.

   - The outside of the ice becomes wet as it starts to melt.
   - If you lift the ice out of the cup by the stick, you will see drops of water coming off the ice.
   - As the ice gets smaller, the liquid water in the cup increases.
   - When there is enough water, the ice floats in the water.
   - The ice gets smaller and smaller as more and more water forms in the cup.
   - Eventually, the last bit of ice slips from the stick.

9. **Observe the melted ice**
   It may take 3 or 4 hours for the ice to completely melt. Gather students, and ask them what would happen if they put the cups of water back in the freezer. Consider doing this another time so that students can reconfirm the process.

10. Review vocabulary

Review the key vocabulary added to the word wall. Here's a suggested cloze review. Students answer chorally.

freeze
ice
liquid
melt
solid

➤ *When water changes to ice , it becomes a _____ .*

S: Solid.

➤ *When ice changes to water, it becomes a _____ .*

S: Liquid.

➤ *To change water from liquid to solid, we _____ .*

S: Freeze it.

➤ *To change water from solid to liquid, we take it out of the cold and let it _____ .*

S: Melt.

11. Answer the focus question

Restate the focus question.

➤ *How can we change water?*

Distribute a strip of paper with the focus question for students to glue into their notebooks. Ask students to answer the focus question in drawings and/or words.

FOCUS CHART

How can we change water?

In the freezer, _____ water changed into _____ water.

WRAP-UP/WARM-UP

12. Share notebook entries

Conclude Part 3 or start Part 4 by having students share notebook entries. Ask students to open their science notebooks to the most recent entry.

Read the focus question together as a class.

➤ *How can we change water?*

Ask students to pair up with a partner to

- share their answers to the focus question;
- explain their drawings.

TEACHING NOTE

*See the **Home/School Connection** for Investigation 5 at the end of the Interdisciplinary Extensions section. This is a good time to send it home with students.*

MATERIALS *for*
Part 4: *Reuse and Recycle Resources*

For each student

- 1   Magnet
- 1   *FOSS Science Resources: Materials in Our World*
  - •   "Land, Air, and Water"

For the class

- •   Bins or basins of reusable or recyclable materials
  - •   Newspaper ★
  - •   Scrap paper from class, junk mail ★
  - •   Empty aluminum cans ★
  - •   Empty steel cans ★
  - •   Empty cardboard boxes ★
  - •   Empty plastic bottles ★
  - •   Small pieces of scrap wood ★
  - •   Fabric scraps ★
  - •   Clean rags ★
- 1   Big book, *FOSS Science Resources: Materials in Our World*

For assessment

- •   *Assessment Checklist*

★ Supplied by the teacher.

GETTING READY *for*

Part 4: *Reuse and Recycle Resources*

1. **Schedule the investigation**

   This part is a whole-class investigation. Plan 15 minutes for the reading and 15 minutes for the recycling activity. Plan 15 minutes for recording in notebooks and to wrap up.

2. **Preview Part 4**

   Students are introduced to natural resources and the need to reuse and recycle materials. They sort materials for recycling, based on the kind of material. Students use magnets to sort steel from other metals. The focus question is **How can we conserve natural resources?**

3. **Organize materials for recycling**

   Organize materials you began collecting in Part 2 that can be reused or recycled. Try to obtain steel cans and aluminum cans so that students can use magnets to separate them. Also bring in a variety of different plastic containers, some glass bottles, and an array of paper products, including newspaper and cardboard boxes.

4. **Find our about local recycling programs**

   Check on your school and community recycling programs, and be ready to tell students about them.

5. **Plan to read *Science Resources*: "Land, Air, and Water"**

   Plan to read "Land, Air, and Water" at the beginning of this part.

6. **Plan assessment**

   Here are specific objectives to observe in this part.

   - *Identify resources from Earth, such as trees, water, land, and air.*

   - *Natural resources are used in everyday life.*

   - *Many resources can be reused and recycled so that they can be conserved.*

   Focus on a few students each session. Record the date and a + or − on the *Assessment Checklist*.

FOCUS QUESTION

How can we conserve natural resources?

Land, Air, and Water

Do you ever **recycle** paper? If you do, you are helping to save a tree.

53

Say it
Write it
New Word
See it
Hear it

Materials for Step 4
- *Materials to recycle*

GUIDING *the Investigation*
Part 4: *Reuse and Recycle Resources*

1. **Read** *Science Resources*: **"Land, Air, and Water"**
   Ask students if they have ever recycled paper. Encourage them to think about the special trash bin in class where they throw out their used paper or the separate place their families put old newspapers, plastic bottles, or aluminum cans. Ask students to think about why people recycle paper. Briefly discuss students' answers. Introduce "Land, Air, and Water," and explain that this article will tell students about other things that need to be taken care of.

   Read aloud "Land, Air, and Water." Pause to discuss key points in the article and to review the pictures.

2. **Discuss the reading**
   Discuss the reading, using these questions as a guide.

   ➤ *Why do we **recycle** paper?*

   ➤ *What do we need land for?*

   ➤ *What do we need clean water for?*

   ➤ *What do we need clean air for?*

3. **Focus question: How can we conserve natural resources?**
   Write the focus question on the chart, and read it together.

   ➤ *How can we **conserve natural resources**?*

   Tell students that things that come from Earth that we use are called natural resources. Point out that land, water, air, and trees are just some of the natural resources on Earth. Some natural resources have properties that allow us to **reuse** them. Others need to be conserved. Explain that *conserve* means to save.

4. **Sort the materials for recycling**
   Show students the bins and basins containing materials that were made from natural resources. Identify the source from which each material is made. Students will know that wood and paper come from trees. All other materials come from materials found in the earth (iron, aluminum, sand, coal, and oil).

   Tell students that they need to sort the different kinds of materials—paper of different kinds, plastic (coal and oil), aluminum, steel (rock), and glass (sand). Work with students to determine how many groups to use to sort the materials. For example, students may want to separate out the different kinds of paper.

5. **Introduce magnets**

   Tell students that **magnets** may help separate the different kinds of cans. See if students can determine how to use the magnets. Pull out several aluminum cans and several steel cans, and let a few students test the cans. See if students can determine that some types of cans stick to the magnet. Ask,

   ➤ *What kinds of cans stick to a magnet?* [Steel cans.]

   Other cans (such as aluminum soft-drink cans) don't stick to magnets. Students can sort the types of cans for recycling, using a magnet.

6. **Set up the class recycling center**

   Set up an area in the room for the recycling center. During the day, let a few students go to the center and sort the materials.

7. **Share community recycling information**

   Review what materials are recycled at your school. Review the community recycling program, and discuss how students can recycle at home.

8. **Review vocabulary**

   Add new words to the word wall, and review vocabulary that was introduced and used during the activity.

9. **Answer the focus question**

   Restate the focus question.

   ➤ *How can we conserve natural resources?*

   Distribute a strip of paper with the focus question for students to glue into their notebooks. Ask students to answer the focus question in drawings and/or words.

Materials for Step 5
- *Magnets*

Say it
New Word
Write it
See it
Hear it

conserve
magnet
natural resources
recycle
reuse

FOCUS CHART

How can we conserve natural resources?

We can recycle _____ to save natural resources.

WRAP-UP/WARM-UP

10. Share notebook entries

Conclude Part 4 or start Part 5 by having students share notebook entries. Ask students to open their science notebooks to the most recent entry. Read the focus question together as a class.

➤ *How can we conserve natural resources?*

Ask students to pair up with a partner to

- share their answers to the focus question;
- explain their drawings.

MATERIALS *for*
Part 5: *Making Sculptures*

For each student

- 10   Craft sticks
- 1   Paper plate
- 1   *FOSS Science Resources: Materials in Our World*
  - •   "I Am Wood"

For the class

- •   Paper scraps (from previous activities) ★
- •   Wood scraps, small ★
- •   Fabric scraps, small ★
- •   Scissors ★
- •   White glue ★
- •   Newspaper ★
- •   Mat–board scraps (optional) ★
- 1   Big book, *FOSS Science Stories: Materials in Our World*
- ❏   1   Teacher master 37, *Center Instructions—Making Sculptures*

For assessment

- •   *Assessment Checklist*

★ Supplied by the teacher.   ❏ Use the duplication master to make copies.

No. 37—Teacher Master

GETTING READY *for*
Part 5: *Making Sculptures*

1. **Schedule the investigation**
   Plan 5–10 minutes to introduce this part to the whole class. Work with groups of six to ten students at a center for 20–40 minutes. Plan to give students as much time as they need, and rotate new students to the center as others finish. Plan 15 minutes for the reading and another 15 minutes for recording in notebooks and to wrap up.

2. **Preview Part 5**
   Students fashion a sculpture, using wood, paper, fabric, glue, and natural materials. Students use the knowledge they have gained in the previous parts of the module to make their own creations. The focus question is **What can we make from different materials?**

3. **Collect wood, paper, and fabric scraps to reuse**
   The *Letter to Family* asks for contributions of wood, paper, and fabric scraps to be sent to school. Collect a variety of different kinds of paper for students to use on their sculptures—wrapping paper, magazines, crepe paper, wallpaper samples, and so on.

   Mat board is very stiff cardboard and is great for building sturdy structures. Picture-framing shops will give you mat-board scraps. Good sources for scrap wood are lumberyards, cabinet shops, and high school wood shops.

4. **Plan glue distribution**
   Students can use a puddle of glue and their fingers, as they did when they made plywood. If you have squeeze bottles of glue, students can apply glue directly to the wood.

5. **Plan a place to dry the sculptures**
   Clear a counter or other fairly large area where the sculptures can
   sit overnight to dry thoroughly.

6. **Set up the center**
   Put the craft sticks in the center of the table. Have a newspaper
   mat and a paper plate ready at each student's place. Paper scraps
   can go in the center of the table or in a central location. Have glue
   and scissors on hand to pass out when students are ready.

7. **Plan to sing "I Am Wood"**
   Plan to sing the song "I Am Wood" included in *Science Resources*.

What can we make from different materials?

Materials for Step 3

- *Samples of available materials to be reused for sculptures*

Materials for Step 4

- *Craft sticks*
- *Scrap paper, fabric, and wood*
- *Paper plates*
- *Newspaper mats*
- *Mat-board scraps (optional)*

GUIDING *the Investigation*
Part 5: *Making Sculptures*

1. **Introduce** *sculpture*
   Call students to the rug. Tell them,

   Today we are going to make **sculptures**. *A sculpture is an artistic creation that looks like something, or reminds you of something, or gives you a feeling. Artists can also be scientists. Artists use what they know about materials to create something. We are going to reuse materials that we have gathered for the sculptures.*

   Write "sculpture" on the word wall.

2. **Focus question: What can we make from different materials?**
   Write the focus question on the chart, and read it together.

   ➤ *What can we make from different materials?*

3. **Describe the materials**
   Show students the paper, fabric, and wood scraps that are available to reuse. Each student will get ten craft sticks to make a sculpture and a paper plate to hold the sculpture. Ask students about the properties of the materials, discussing ways they can be used to create their works of art. Paper strips can be curled, woven, torn, or folded; wood can be glued and painted; and fabric can be cut and folded.

   Encourage students to keep their sculptures on the paper plates so that their work will be easy to move to a drying spot when they are finished.

4. **Send students to the center**
   Send six to ten students to the center. Let them choose ten craft sticks from the center of the table and select pieces of scrap paper, fabric, and wood for their sculptures.

5. **Start designing**

Let students work freely for a time before glue is provided. Tell them to pile and hold pieces of wood, fabric, and paper together in various ways to figure out how they would like to assemble their sculptures.

If students have difficulty getting started, make some suggestions that include two or more materials. Students may be interested in making boats with sails; people with masks or hats; animals with paper tails, fur, or wings; houses with paper doors, windows, and flags; or an abstract sculpture.

6. **Assemble the sculptures**

Distribute the glue and scissors, and let the creating begin. Encourage students to keep their sculptures on their paper plates if possible.

7. **Label the sculptures**

As students continue to work, write each student's name and the date on a craft stick. Students should include the craft sticks as part of their sculptures so they can later identify their work.

8. **Dry the sculptures overnight**

As each student finishes, set his or her sculpture in a place where it can dry overnight. Check to make sure it is labeled with the student's name.

9. **Share constructions**

When students have finished, let each one have a chance to share what he or she created. Students should also explain how they made their sculptures, using relative position words such as above, below, behind, in front of, and beside.

Materials for Step 6
- *Scissors*
- *White glue*

READING *in Science Resources*

10. Sing "I Am Wood"

Students have had multiple opportunities to investigate the properties and uses of wood. This song reinforces students' learning by discussing how wood is used, how wood feels, and why people need to replant trees—the natural resource that wood comes from.

Ask students how it would feel to be wood. Briefly discuss answers. Continue the discussion by asking,

➤ *If you were wood, what item would you be—a tree, a chair, a table?*

Introduce the song "I Am Wood," and tell students that they will be learning a new song about wood. Ask students to listen carefully so that they can learn the song and understand what it is saying.

Read the song aloud to students, modeling the melody. Continue to teach the song, using an echo-singing approach. Practice and sing the song as it fits into your schedule.

11. Discuss the song

Discuss the song, using this question as a guide.

➤ *What is the song saying about wood?*

NOTE: A recording of "I Am Wood" can be found on FOSSweb.

INTERDISCIPLINARY EXTENSIONS

Science Extensions

TEACHING NOTE

Review the teacher resources on FOSSweb for community- and career-related extensions.

- **Finish the sculptures**
  When the sculptures are dry, let students "stain" the wood with brown tempera paint. Brown is a standard color for a stain, but any color of the rainbow can be used to create a dramatic or an offbeat design. Here's how to "stain" the wood.

  a. Brush the stain (tempera paint) on the wood, using long, even strokes.

  b. Allow the stain to soak in for a minute or so.

  c. Dip a paper towel in water, and use it to wipe off as much stain as possible.

  d. Finish by rubbing the wood with a dry paper towel until no more stain comes off.

- **Maintain a workbench**
  If some students take a shine to woodworking, set up a permanent woodworking center in your classroom.

- **Set up a school recycling center**
  Have students work with the school administrator and the custodian to set up a recycling center at school for paper, plastic, and metal.

TEACHING NOTE

Families can get more information about Home/School Connections from FOSSweb.

Home/School Connection

At home, students can gain more experiences with freezing and melting water.

Make copies of teacher master 38, *Home/School Connection* for Investigation 5, and send them home with students after Part 3.

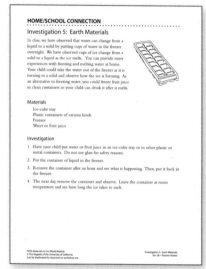

No. 38—Teacher Master

Assessment

THE FOSS ASSESSMENT SYSTEM FOR KINDERGARTEN

Contents

The most important FOSS goal is scientific literacy for students. Scientific literacy is the ability to do, understand, and communicate science. Achieving this goal entails a great deal of student learning. It is critically important to make the quality and extent of that learning visible. Knowing what students know is the key to a highly effective science program. That's what the FOSS assessment system is designed to find out.

The FOSS assessment system includes both formative and summative assessments. Formative assessment looks at learning during the process of instruction. It measures progress, provides information about learning, and is generally diagnostic. Summative assessment looks at the learning after instruction is completed. It measures achievement and is generally evaluative.

Formative assessment in FOSS, called **embedded assessment**, occurs on a daily basis. Teachers observe action in class or review notebooks after class. Embedded assessments provide continuous monitoring of students' learning and help you make decisions about whether to review, extend, or move on to the next idea to be covered.

The FOSS assessment system has another component for older students. **Benchmark assessments** are short summative assessments given after each investigation. These **I-Checks** are actually hybrid tools that provide summative information about student achievement, but because they occur soon after teaching each investigation, they can be used diagnostically as well. Reviewing a specific item on an I-Check with the class provides another opportunity for students to clarify their thinking.

For the kindergarten modules, assessment is exclusively embedded—you observe students' actions in class.

EMBEDDED *Assessment*

Assessment and teaching go hand in hand. Assessing students on a regular basis gives you valuable information to guide instruction and to keep families and other interested members of the educational community informed about students' progress. Assessment should be an ongoing part of everyday life in the classroom.

FOSS embedded assessments allow you and your students to monitor learning on a daily basis as you progress through the module. You will find suggestions for what to assess in the Getting Ready section of each part of each investigation. For example, here is a typical Getting Ready step for Part 1 of the first investigation.

---

14. Plan assessment for Part 1

There are six objectives that can be assessed at any time during any part of this investigation.

What to Look For

- *Students ask questions.*

- *Students plan and conduct simple investigations.*

- *Students use senses to observe materials and objects.*

- *Students record and organize observations.*

- *Students communicate observations orally and in their notebooks with words and drawings.*

- *Students incorporate new vocabulary.*

Here are the specific content objectives to observe in this part.

- *Wood has observable properties and can be described by those properties.*

- *Wood is a resource that comes from trees.*

Here are the specific scientific practices to observe in this part.

- *Students compare properties of wood (organize observations).*

- *Students sort wood by properties (organize observations).*

A two-page *Assessment Checklist* is provided in the Assessment Masters to record observations about students' progress at the science center or in students' notebooks. Photocopy the pages, and attach them to a clipboard so you can keep them nearby.

Assessment Checklist

The first page of the *Assessment Checklist* is for content knowledge dealing with concepts—matter has structure, and matter interacts—developed throughout the module. You can record on one sheet for all five investigations, or make a copy of the sheet and use one sheet for each investigation. There is space in the title to record the investigation number.

The second page is a checklist for the scientific and engineering practices. There is space on this sheet for writing notes about each student.

Observing at the learning center. You will find that opportunities for students to develop the concepts and skills described above abound in the science center. The Getting Ready section for each part highlights those assessments in a particular activity. Focus on a few students at a time when using the checklist. Over the course of the module, you'll observe each student several times. Each time you observe a student, mark the *Assessment Checklist* with the date and a + or – to indicate progress. This system will also help you keep track of how often you have observed each student. Look for improvement over the course of the module.

Science notebooks. Making good observations and using them to develop explanations for how the natural world works is the essence of science. This process calls for critical thinking and good communication skills. Students' notebooks are a useful assessment tool and a good language extension. Use the notebook entries to provide you with evidence that students are accomplishing the learning goals of the module.

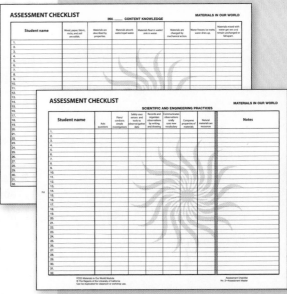

Nos. 1–2—Assessment Masters

Narrative Report

The Assessment Masters include an individual *Narrative Report* to send home to families or pass on to first-grade teachers. Make one copy of this two-page report for each student, and fill it out, using the records you have been keeping on the *Assessment Checklist*. In the comments section, write a few observations about the progress you have seen in each student's science learning from the beginning to the end of the module.

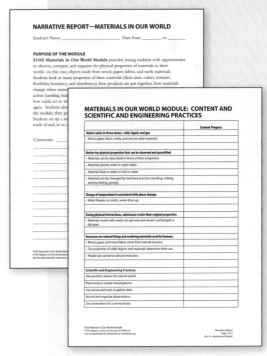

Nos. 3–4—Assessment Masters

SUMMARY *of Assessment* Opportunities in Kindergarten

	Inv. 1	Inv. 2	Inv. 3	Inv. 4	Inv. 5
Matter Has Structure					
A Matter exists in three states: solid, liquid, and gas.					
• Wood, paper, fabric, rocks, and soil are solid materials.	✓	✓	✓	✓	✓
B Matter has physical properties that can be observed and quantified.					
• Materials can be described in terms of their properties.	✓	✓	✓	✓	✓
• Materials absorb water or repel water.	✓	✓	✓	✓	✓
• Materials float in water or sink in water.	✓	✓			✓
• Materials can be changed by mechanical action (sanding, cutting, tearing, folding, gluing).		✓	✓	✓	
Matter Interacts					
B Change of temperature is associated with phase change.					
• Water freezes, ice melts, water dries up.		✓		✓	✓
C During physical interactions, substances retain their original properties.					
• Materials mixed with water can get wet and remain unchanged or fall apart.		✓	✓	✓	✓
Scientific and Engineering Practices					
Ask questions about the natural world.	✓	✓	✓	✓	✓
Plan/conduct simple investigations.	✓	✓	✓	✓	✓
Use senses and tools to gather data.	✓	✓	✓	✓	✓
Record and organize observations.	✓	✓	✓	✓	✓
Use observations to communicate.	✓	✓	✓	✓	✓
Natural resources are used by people.					
• Wood, paper, and most fabric come from natural sources.	✓	✓	✓	✓	
• The properties of solid objects and materials determine their use.	✓	✓	✓	✓	✓
• People can conserve natural resources.			✓		✓